Continuous Testing, Quality, Security, and Feedback

Essential strategies and secure practices for DevOps, DevSecOps, and SRE transformations

Marc Hornbeek

Continuous Testing, Quality, Security, and Feedback

Group Product Manager: Preet Ahuja
Publishing Product Manager: Surbhi Suman
Book Project Manager: Srinidhi Ram
Senior Editor: Sayali Pingale
Technical Editor: Arjun Varma
Copy Editor: Safis Editing
Indexer: Tejal Soni
Production Designer: Ponraj Dhandapani
DevRel Marketing Coordinator: Rohan Dobhal

First published: August 2024

Production reference: 1080824

Published by Packt Publishing Ltd.

Grosvenor House
11 St Paul's Square
Birmingham
B3 1RB, UK

ISBN 978-1-83546-224-9

www.packtpub.com

To my wife, Virginia, for giving me the motivation to continue my life and work. To my sister, Christine, for her support. To the memory of my loving parents.

– Marc Hornbeek

Foreword

I am honored to write the foreword for this book by Marc Hornbeek, a leading expert and practitioner of continuous testing, quality, security, and feedback. I have worked with Marc to adopt continuous quality into our CI/CD delivery pipelines, and I can attest to his deep knowledge and practical experience in this field. His expertise and guidance were instrumental in our success in delivering software that met our customers' expectations and needs. The success of the transformative project was recognized with the Best DevOps Industry Implementation DevOps Dozen Community award in 2022.

Marc shares his insights and best practices for mastering the strategies and secure practices for DevSecOps and SRE transformations in this book. He explains how to build automation that addresses the shift toward more continuous, integrated, automated, and user-focused practices to deliver high-quality and secure software faster and with greater reliability. He covers the key concepts, principles, tools, and techniques that enable continuous testing, measurement, quality, security, and feedback throughout the software development lifecycle. He also provides real-world examples and case studies that illustrate applying these practices in different contexts and scenarios.

As companies like ours race to move toward continuous production deployments at high frequency, building in continuous feedback loops and automated quality gates that provide essential guidance to our developers and other staff is crucial to ensure our code is delivered quickly and is secure and of high quality. This book will help you understand how to achieve this goal and overcome the common challenges and pitfalls you may encounter along the way. Whether you are a developer, tester, engineer, manager, SRE, or senior leader, you will find valuable information and advice in this book that will help you improve your software delivery processes and outcomes.

I highly recommend this book to anyone interested in learning about continuous testing, measurement, quality, security, and feedback and how to implement them in your organization. Marc has done a great job of distilling his vast experience and wisdom into a concise and comprehensive guide that will benefit the readers. I hope you enjoy reading this book as much as I did and that you find it valuable and inspiring for your journey toward continuous quality delivered at the rate your customers demand.

Dan Wakeman

SVP Development of Engineering Excellence, FIS

Contributors

About the author

Marc Hornbeek, a.k.a *DevOps-the-Gray*, is the CEO and principal consultant at Engineering DevOps Consulting. He is the author of the book *Engineering DevOps*, and serves as an ambassador, author, and instructor for the DevOps Institute. Marc also blogs on websites such as DevOps, CloudNativeNow, and SecurityBoulevard.

Globally recognized as a strategic consultant, Marc applies engineering practices holistically for continuous testing, DevOps, DevSecOps, and SRE digital transformations. He has led more than 90 transformations for enterprises, manufacturers, service providers, and government institutions.

Marc is an IEEE Outstanding Engineer and an IEEE Life Member. His education includes degrees in engineering and executive business, as well as multiple certifications from the DevOps Institute.

About the reviewers

Debashis Bhattacharyya has worked in the tech industry for over 18 years. He has planned, architected, designed, and built multiple technology solutions over the years. He specializes in cloud, API, data, DevSecOps, digital transformation, and payment application modernization. He has an engineering degree from Anna University. He has worked on large- and medium-scale transformation projects for multiple companies designing and building DevSecOps processes. He has written articles and white papers and has been featured in podcasts and webinars discussing DevSecOps. He also led the tech delivery of the team that won the DevOps Dozen Award from the DevOps Institute for Best DevOps Industry Implementation in 2022.

It takes a lot of time and commitment to read, research, and review a book on a topic that is constantly evolving. Hence, I'd like to thank my wife, Soundharya (Sandy), kids, Rihaan and Ved, and pet, Gucci, for understanding and giving me the space and time to work on this project during after-office hours, which are otherwise dedicated to them.

I would also thank my colleagues for making work so much fun, and my friends for always being there for me.

Victorio Mosso is the founder of ANALYTICA MTY. He has developed his career in the IT service management industry for more than 18 years. He has participated in diverse areas in global organizations such as software development, service support, and service delivery, data and performance management, and DevOps. He achieved the ITIL Master designation and he has been designated as ITIL and a DevOps ambassador.

I'd like to thank Marc for the honor of being part of this great project. His emphasis on achieving excellence through an engineering approach is truly amazing and inspiring. I am also thankful to my family for motivating me to keep learning every day.

Chetan Talwar is a solution architect specializing in architecting scalable solutions. With extensive experience delivering workshops and speaking at industry events, he focuses on cloud computing, DevOps, and automation. He excels at translating complex technical concepts into practical applications, helping businesses optimize their cloud infrastructure. His expertise ensures efficient, resilient, and secure solutions by integrating automation and DevOps practices. As a passionate educator, he shares insights at conferences, emphasizing the future of cloud computing and the role of automation in driving business growth and efficiency.

Table of Contents

3

Experiences and Pitfalls with Continuous Testing, Quality, Security, and Feedback   39

Part 2: Determining Solutions Priorities

4

Engineering Approach to Continuous Testing, Quality, Security, and Feedback   59

5

Determining Transformation Goals   81

6

Discovery and Benchmarking   109

Part 3: Deep Dive into Roadmaps, Implementation Patterns, and Measurements

9

10

11

Understanding Transformation Implementation Patterns   239

12

Measuring Progress and Outcomes   257

Part 4: Exploring Future Trends and Continuous Learning

13

14

Preface

In the rapidly evolving landscape of software development, the integration of continuous testing, quality, security, and feedback has become pivotal for organizations aiming to achieve successful digital transformations. *Continuous Testing, Quality, Security, and Feedback* is a comprehensive guide that delves into the core strategies necessary for embedding these practices into the heart of DevOps, DevSecOps, and SRE methodologies.

The book begins by setting the stage for understanding the critical role of continuous testing, quality, security, and feedback in the context of digital transformations. It provides a historical perspective, illustrating how these strategies have evolved from traditional approaches to become integral components of Agile, DevOps, DevSecOps, and SRE practices. This foundational knowledge is crucial for professionals to appreciate the necessity of integrating these continuous strategies into their workflows to enhance speed, efficiency, and reliability in software delivery.

One of the book's strengths lies in its clear, outcome-focused definitions of continuous testing, quality, security, and feedback. These definitions guide professionals in implementing these strategies effectively within their organizations. By aligning these practices with measurable business outcomes, particularly those recognized by the **DevOps Research Association** (**DORA**), the book ensures that you can evaluate and adjust their methodologies based on their impact on key performance indicators. This approach not only provides clarity but also emphasizes the importance of focusing on results rather than merely procedural actions.

The core of the book is dedicated to exploring the guiding principles and pillars that underpin continuous testing, quality, security, and feedback. Through detailed exposition, you will be equipped with the knowledge to integrate testing into every stage of the software development life cycle, adopt a proactive approach to quality and security, and foster a culture of continuous feedback and improvement. These sections are invaluable, offering practical insights and strategies for overcoming common challenges and leveraging best practices to achieve high-quality, secure, and user-centric software products.

The book is more than just a theoretical guide; it is a catalyst for transformation. It encourages professionals to embrace continuous strategies, ensuring that digital transformations are resilient, user-centric, and secure.

Who this book is for

Whether you are a seasoned expert or a newcomer to the field, this book provides valuable insights and skills that will elevate your approach to continuous software development, delivery, and operations. This book is an essential resource for anyone looking to implement or enhance continuous testing, quality, security, and feedback within their DevOps, DevSecOps, and SRE practices. It offers a practical guide and a comprehensive framework for achieving efficiency, reliability, and success in digital transformations, making it a must-read for professionals committed to excellence in software development and operations.

What this book covers

Chapter 1, Principles of Continuous Testing, Quality, Security, and Feedback, explains how these strategies are essential for digital transformations that utilize continuous development practices known as Agile, continuous delivery practices known as DevOps and DevSecOps, and continuous operations practices known as SRE.

Chapter 2, The Importance of Continuous Testing, Quality, Security, and Feedback, explains why continuous testing, quality, security, and feedback strategies are important for DevOps, DevSecOps, and SRE. It explains how the principles and pillars of DevOps, DevSecOps, and SRE depend on the principles and pillars of continuous testing, quality, security, and feedback.

Chapter 3, Experiences and Pitfalls with Continuous Testing, Quality, Security, and Feedback, explains – by way of examples from my experiences – use cases, lessons learned, and pitfalls to avoid, including strategies to avoid pitfalls.

Chapter 4, Engineering Approach to Continuous Testing, Quality, Security, and Feedback, explains a systematic, disciplined engineering approach to planning continuous testing, quality, security, and feedback solutions.

Chapter 5, Determining Transformation Goals, explains a prescriptive methodology for determining goals for continuous testing, quality, security, and feedback transformations, to suit specific organizations, products, and services. Tools to help determine goals are described.

Chapter 6, Discovery and Benchmarking, explains the methodology and tools for discovering the current state of an organization's people, processes, and technologies relevant to the transformation to mastering continuous testing, quality, security, and feedback.

Chapter 7, Selecting Tool Platforms and Tools, provides you with a deep understanding of how each platform and tool can be leveraged to foster a culture of continuous improvement and resilience in the face of ever-changing technological challenges.

Chapter 8, Applying AL/ML to Continuous Testing, Quality, Security, and Feedback, delves into the transformative role of **Artificial Intelligence** (**AI**) and **Machine Learning** (**ML**) across the software development life cycle, with a special focus on enhancing continuous testing, quality, security, and feedback practices.

Chapter 9, Use Cases for Integrating with DevOps, DevSecOps, and SRE, describes practical applications of continuous testing, continuous quality, continuous security, and continuous feedback within these frameworks with use cases that illustrate how organizations can transform to higher levels of operational maturity.

Chapter 10, Building Roadmaps for Implementation, explains how to create effective roadmaps for implementing continuous testing, quality, security, and feedback within your organization, ensuring that your digital transformation journey is both strategic and aligned with organizational goals.

Chapter 11, Understanding Transformation Implementation Patterns, dives into the world of implementation patterns, which are structured approaches proven to enhance the deployment and success of strategic roadmaps for organizations that wish to improve their capabilities for continuous testing, quality, security, and feedback.

Chapter 12, Measuring Progress and Outcomes, focuses on methods and frameworks that are important for measuring progress and outcomes as organizations implement and improve their continuous testing, quality, security, and feedback capabilities.

Chapter 13, Emerging Trends, describes emerging trends that are reshaping the landscape of continuous testing, quality, security, and feedback within software development.

Chapter 14, Exploring Continuous Learning and Improvement, explains effective strategies for continuous learning and improvement in areas crucial for software development and operations: continuous testing, quality, security, and feedback.

To get the most out of this book

There are no specific code files, tools, or software applications required to understand or use this book. However, there are examples, templates, and tools that are made available to supplement the materials in the book at `https://github.com/PacktPublishing/Continuous-Testing-Quality-Security-and-Feedback`.

> **Note**
> For those interested in the author's consulting services,
> please visit `www.engineeringdevops.com` to get in touch.

Conventions used

There are a number of text conventions used throughout this book.

Bold: Indicates a new term, an important word, or words that you see onscreen. For instance, words in menus or dialog boxes appear in **bold**. Here is an example: "Select **System info** from the **Administration** panel."

> **Tips or important notes**
> Appear like this.

Get in touch

Feedback from our readers is always welcome.

General feedback: If you have questions about any aspect of this book, email us at customercare@ packtpub.com and mention the book title in the subject of your message.

Errata: Although we have taken every care to ensure the accuracy of our content, mistakes do happen. If you have found a mistake in this book, we would be grateful if you would report this to us. Please visit www.packtpub.com/support/errata and fill in the form.

Piracy: If you come across any illegal copies of our works in any form on the internet, we would be grateful if you would provide us with the location address or website name. Please contact us at copyright@packt.com with a link to the material.

If you are interested in becoming an author: If there is a topic that you have expertise in and you are interested in either writing or contributing to a book, please visit authors.packtpub.com.

Share Your Thoughts

Once you've read *Continuous Testing, Quality, Security, and Feedback*, we'd love to hear your thoughts!
Scan the QR code below to go straight to the Amazon review page for this book and share your feedback.

https://packt.link/r/1835462243

Your review is important to us and the tech community and will help us make sure we're delivering
excellent quality content.

Download a free PDF copy of this book

Thanks for purchasing this book!

Do you like to read on the go but are unable to carry your print books everywhere?

Is your eBook purchase not compatible with the device of your choice?

Don't worry, now with every Packt book you get a DRM-free PDF version of that book at no cost.

Read anywhere, any place, on any device. Search, copy, and paste code from your favorite technical books directly into your application.

The perks don't stop there, you can get exclusive access to discounts, newsletters, and great free content in your inbox daily

Follow these simple steps to get the benefits:

1. Scan the QR code or visit the link below

https://packt.link/free-ebook/9781835462249

2. Submit your proof of purchase
3. That's it! We'll send your free PDF and other benefits to your email directly

Part 1: Understanding Continuous Testing, Quality, Security, and Feedback

Part 1 of the book dives into the foundational concepts necessary for integrating continuous strategies into software development and operations. It begins by explaining the principles of continuous testing, quality, security, and feedback, emphasizing their critical role in supporting Agile, DevOps, DevSecOps, and SRE practices. This section sets the stage by outlining the historical context and evolution of these strategies, highlighting how they have become essential in modern software frameworks to enhance efficiency, security, and user responsiveness.

Further, the book discusses the importance of these continuous strategies in maintaining and improving the quality, security, and feedback mechanisms within software development processes. It uses real-world examples and lessons learned from personal past experiences to illustrate common pitfalls and effective strategies to avoid them. This part helps in understanding the theoretical aspects and provides practical insights into implementing these strategies effectively to achieve robust digital transformations.

This part includes the following chapters:

- *Chapter 1, Principles of Continuous Testing, Quality, Security, and Feedback*
- *Chapter 2, The Importance of Continuous Testing, Quality, Security, and Feedback*
- *Chapter 3, Experiences and Pitfalls with Continuous Testing, Quality, Security, and Feedback*

Principles of Continuous Testing, Quality, Security, and Feedback

This chapter explains how the continuous strategies are essential for digital transformations that utilize continuous development practices known as Agile, continuous delivery practices known as DevOps and DevSecOps, and continuous operations practices known as **Site Reliability Engineering** (**SRE**).

In this chapter, we'll cover the following main topics:

- Introducing continuous testing, continuous quality, continuous security, and continuous feedback
- Defining continuous testing, quality, security, and feedback
- The guiding principles and pillars of continuous testing
- The guiding principles and pillars of continuous quality
- The guiding principles and pillars of continuous security
- The guiding principles and pillars of continuous feedback

Let's get started!

Introducing continuous testing, quality, security, and feedback

This section introduces the key foundational concepts and historical context for modern continuous testing, quality, security, and feedback strategies. It also explains why DevOps, DevSecOps, and SRE practices drive the need for continuous testing, quality, security, and feedback.

Foundations for testing, quality, security, and feedback

Testing, quality, security, and feedback have been integral to software development, delivery, and operations since the inception of software. *Figure 1.1* and the following paragraphs depict some historical examples that highlight this.

ENIAC (1940s)   IBM 360 Mainframe (1964)   Morris Worm (1988)

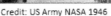

Credit: US Army NASA 1946   Credit: IBM   Credit: WeLiveSecurity

Figure 1.1 – Early examples of testing, quality, security, and feedback

- **Testing and quality – ENIAC (1940s):** Even with the first general-purpose electronic computer, ENIAC, testing and debugging were crucial. The machine had to be meticulously programmed and tested for each new task, a process that often took days. This early example underscores the importance of testing for quality assurance in software.

- **Security – The Morris Worm (1988):** The Morris Worm, one of the first recognized worms to affect the world's nascent internet infrastructure, highlighted the need for attention to security in software design. It exploited known vulnerabilities, which underscored the importance of security in networking and software development.

- **Feedback – IBM's early software development (1950s–1960s):** In the early days of commercial software, institutions and companies such as IBM realized the importance of customer feedback in software development. Feedback from users helped shape the evolution of software products, making them more user-friendly and aligned with business needs.

However, the traditional methods had some drawbacks. Let's look at them next.

The weaknesses of traditional testing, quality, security, and feedback strategies

The historical examples of ENIAC, the Morris Worm, and IBM's early software development highlight key weaknesses in traditional approaches to testing, quality, security, and feedback in software development:

- **Testing and quality – ENIAC (1940s):** With ENIAC, testing and debugging were manual and time-consuming. Each new task required meticulous programming and testing, demonstrating the inefficiency of traditional testing methods in the face of complex tasks. The absence of automated testing tools and integrated testing practices meant that ensuring quality was a labor-intensive process, significantly slowing down development and deployment.

- **Security – The Morris Worm (1988)**: Traditional approaches often treated security as an afterthought. The Morris Worm exploited known vulnerabilities, highlighting the weakness of reactive security measures in contrast to the need for proactive security practices. Security was not integrated into the software development life cycle. The incident underscored the importance of considering security at every stage of development, from design to deployment.

- **Feedback – IBM's early software development (1950s–1960s)**: Traditional software development often suffered from delayed feedback loops. Feedback was typically collected post-release, limiting the ability to make user-centric improvements during the development phase. There was a lack of continuous engagement with users during the development process. Feedback was not systematically integrated into the development cycle, leading to products that might not fully align with user needs or expectations.

These historical examples illustrate key weaknesses in traditional approaches:

- **Testing and quality**: Manual, time-consuming testing methods, lack of automation, and a failure to integrate testing into the development life cycle.

- **Security**: A reactive approach to security, treating it as an afterthought rather than an integral part of the development process.

- **Feedback**: Delayed feedback mechanisms and a lack of continuous user engagement, leading to a disconnect between software development and user requirements.

Now, let's consider how testing, quality, security, and feedback evolved as software frameworks became more continuous.

The evolution of testing, quality, security and feedback toward continuous strategies

The evolution of software development, delivery, and operations toward continuous development, delivery, and operation methodologies such as Agile, DevOps, DevSecOps, and SRE was driven by several key factors and industry trends:

- **An increasing demand for speed and agility**: As markets and technology rapidly evolved, businesses faced growing pressure to deliver products and services faster. This need for speed led to the adoption of Agile methodologies, which focus on iterative development, flexibility, and fast delivery of software.

- **A shift from a project to a product mindset**: Traditional software development was often project-based, with a clear start and end. However, the industry shifted toward a product mindset, where software is continuously developed, improved, and maintained. This ongoing nature of software products necessitated methodologies such as Agile and DevOps.

- **The complexity of modern software systems**: The increasing complexity of software systems, with distributed architectures such as microservices, demanded more collaborative and integrated approaches. DevOps emerged as a response, emphasizing collaboration between **development (Dev)**, **quality (QA)**, **security (Sec)**, and **operations (Ops)** teams.

- **Need for faster release cycles**: With growing competition and technological advancements, the ability to release updates and features quickly became a competitive advantage. This led to the adoption of **Continuous Integration/Continuous Delivery (CI/CD)** practices within DevOps frameworks.

- **The rise of cloud computing and automation**: The advent of cloud computing and the increasing availability of automation tools allowed for more efficient and scalable software development, delivery, and operations processes. These technologies are fundamental to DevOps, DevSecOps, and SRE practices.

- **Growing importance of security**: With the rise in cyber threats and security breaches, integrating security into the software development life cycle became crucial. DevSecOps evolved from DevOps by incorporating Sec as a key component from the outset of development projects.

- **Focus on reliability and user experience**: As user expectations for reliability and performance grew, there was a shift in focus toward ensuring that software is not only delivered quickly but also reliably. This led to the emergence of SRE, which blends aspects of software engineering with IT operations to create scalable and highly reliable software systems.

- **Feedback and continuous improvement**: The need for continuous feedback from users and rapid adaptation to this feedback became paramount. Agile, DevOps, and SRE methodologies all emphasize continuous monitoring, feedback, and improvement to align software products more closely with user needs and business goals.

- **Cultural and organizational shifts**: These methodologies also represent a cultural shift in how organizations view software development, delivery, and operations. They promote collaborative, cross-functional teams, a fail-fast mindset, and an emphasis on continuous learning and improvement.

The evolution to Agile, DevOps, DevSecOps, and SRE has been driven by the need for faster, more efficient, and more reliable software delivery in a rapidly changing technological landscape. These methodologies address the increasing complexity of software systems, the need for speed and reliability, the integration of security into the development process, and the importance of continuous feedback and improvement.

The historical examples presented in this section demonstrate that, from the earliest days of computing, strategies for testing, quality, security, and feedback have been critical components of software development, delivery, and operations. These strategies have evolved with the evolution of technology but have always been integral to the development, delivery, and maintenance of reliable, secure, and user-centered software.

Weaknesses of traditional strategies for testing, quality, security, and feedback led to the evolution of more integrated, automated, and user-centric methodologies in software development, such as Agile, DevOps, and DevSecOps, which aim to address these shortcomings by embedding testing, quality assurance, security, and feedback deeply and continuously into the software life cycle.

The next section will explain how original strategies for testing, quality, security, and feedback have evolved to keep pace with the modern era of continuous development, delivery, and operations.

Evolution toward continuous testing, quality, security, and feedback

The advent of Agile, DevOps, DevSecOps, and SRE practices has necessitated a significant evolution in the testing, quality, security, and feedback strategies. This evolution, illustrated in *Figure 1.2*, is driven by changes in technology, business needs, and the continuous approach to software development, delivery, and operations.

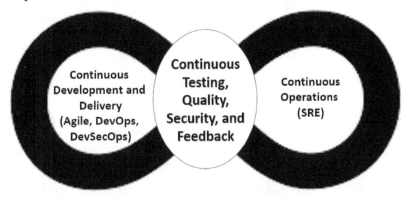

Figure 1.2 – Continuous testing, quality, security, and feedback

Let's explore the specific attributes of testing, quality, security, and feedback strategies that needed to evolve to become "continuous":

- **Testing in the context of Agile, DevOps, DevSecOps, and SRE:**

  - **Faster release cycles**: Traditional testing methods were too slow for the rapid deployment cycles in DevOps. CI/CD pipelines required automated, more frequent, and more sophisticated testing strategies to ensure that new features and updates could be deployed quickly without compromising quality.

  - **Shift-left testing**: DevOps advocates for *shifting left* in the software development life cycle, meaning testing begins much earlier in the process. This shift ensures that defects are caught and addressed sooner, reducing costs and time-to-market.

- **Reliability and availability**: In SRE, the focus is on the reliability, availability, and performance of services. Testing here includes not just functional testing but also load, performance, and resilience testing to ensure the system can handle real-world scenarios.

- **Quality in the era of Agile, DevOps, DevSecOps, and SRE**:

  - **A user-centric focus**: The rapid and iterative nature of DevOps requires a user-centric approach to quality. Features and updates are rolled out continually, and the quality of these increments directly impacts user experience.

  - **Monitoring and performance metrics**: SRE places significant emphasis on using real-time monitoring and performance metrics to maintain and improve the quality of services. These metrics are vital for making data-driven decisions about system improvements.

- **Security in the context of Agile, DevOps, DevSecOps, and SRE**:

  - **Continuous security**: The traditional model of addressing security at the end of the software development life cycle is not viable in a fast-paced DevOps environment. DevSecOps integrates security into every stage of the development process, while SecOps integrates security protections in products; together, they ensure continuous security.

  - **Automated security testing**: Automation in security testing is crucial in DevSecOps and SecOps. Tools that automatically scan original and third-party code for vulnerabilities are integrated into the CI/CD pipeline, allowing you to immediately detect and remediate security issues. Also included is the automation of penetration testing and tools that monitor and protect operating software in production to identify and mitigate security intrusions, enabling an improved defense of deployed software systems in production.

  - **Compliance as code**: In DevSecOps, compliance requirements are codified so that they can be automatically and consistently enforced throughout the development and operations life cycle.

- **Feedback in the context of Agile, DevOps, DevSecOps, and SRE**:

  - **A continuous feedback loop**: DevOps, DevSecOps, and SRE practices thrive on a continuous feedback loop between development, operations, and the user. This feedback is crucial for the rapid iteration of software delivery and deployment to production operations.

  - **Blameless postmortems**: SRE practices such as conducting blameless postmortems after incidents facilitate a culture of learning and improvement. This approach allows teams to understand what went wrong and how to prevent it in the future, without focusing on individual faults.

  - **Cross-functional collaboration**: Feedback in these methodologies is not just about user input but also involves cross-functional team collaboration. Sharing insights and knowledge between development, operations, security, and other stakeholders is key to improving processes and outcomes.

Figure 1.3 illustrates the relationships between continuous testing, quality, security, and feedback relative to continuous development (Agile), continuous delivery (DevOps and DevSecOps), and continuous operations (SRE). The figure shows that continuous testing, quality, and security are active during the development, delivery, and operations phases. It also shows that results from each phase, resulting from these strategies, provide continuous feedback data that affects the continuous iterations of each phase.

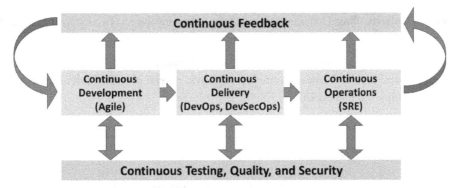

Figure 1.3 – The continuous testing, quality, security, and feedback relationships

The evolution of testing, quality, security, and feedback in the context of DevOps, DevSecOps, and SRE reflects a broader shift in software development and operations. This shift is toward more continuous, integrated, automated, and user-focused practices, aimed at delivering high-quality, secure software at a faster pace and with greater reliability.

Defining continuous testing, quality, security, and feedback

This section explains the importance of defining continuous testing, quality, security, and feedback and the challenges associated with doing so, followed by practical definitions for continuous testing, quality, security, and feedback.

The need for definitions of testing, quality, security, and feedback

There are no standard definitions for continuous testing, quality, security, and feedback, just as there are no standard definitions for DevOps, DevOps, or SRE. However, defining continuous testing, quality, security, and feedback within the context of an organization's transformation to mature DevOps, DevSecOps, and SRE practices is crucial for several reasons. Definitions provide a foundation to establish metrics and measure the performance and progress for people, processes, and technologies dimensions of holistic digital transformations. The importance, potential benefits, and consequences of having (or not having) these clear definitions are explained in this section.

The importance of definitions

Let's understand why definitions are important:

- **A basis for measurement**: Clear definitions allow organizations to establish specific, measurable criteria to evaluate the effectiveness of their practices. This is essential for continuous improvement.

- **Common understanding**: Definitions ensure that everyone involved has a common understanding of what is expected, reducing ambiguities and misalignments across teams.

- **Goal alignment**: Well-defined concepts help align the goals of various teams (development, operations, and security) toward a unified objective, crucial in collaborative environments such as DevOps and DevSecOps.

The benefits of clear definitions

Let's look at the benefits of definitions:

- **Performance tracking**: With clear definitions, organizations can track the performance of their DevOps, DevSecOps, or SRE initiatives over time, identifying areas of success and those needing improvement.

- **Improved collaboration**: Definitions facilitate better communication and collaboration between teams, as everyone operates with a shared understanding of key concepts.

- **Focused training and development**: Definitions enable targeted training and development efforts, focusing on specific areas identified through these definitions and metrics.

- **Enhanced process optimization**: Organizations can more effectively identify and implement process optimizations, leading to increased efficiency, reduced costs, and higher-quality output.

The consequences of a lack of clear definitions

Next, we'll understand what happens when objectives are not clearly defined:

- **Measurement challenges**: Without clear definitions, it becomes challenging to measure and assess the effectiveness of DevOps, DevSecOps, and SRE practices, leading to potential inefficiencies and unaddressed problems.

- **Misaligned goals**: Ambiguities can lead to misaligned goals and expectations among teams, resulting in conflicts and reduced synergy.

- **Ineffective resource allocation**: Unclear definitions make it difficult to identify where resources should be allocated for maximum impact, potentially leading to wasted effort and investment.

- **Reduced accountability**: It becomes harder to hold teams and individuals accountable for their roles and responsibilities in the absence of well-defined criteria for success.

Clear and unambiguous definitions of continuous testing, quality, security, and feedback provide the necessary groundwork to set and measure performance metrics, ensuring everyone is aligned toward common goals and facilitating continuous improvement. The lack of such definitions can hinder the progress toward maturing practices in these areas.

The challenges of defining continuous testing, quality, security, and feedback

Standardizing the definitions of continuous testing, quality, security, and feedback is a challenging task for any organization, due to the dynamic and varied nature of software development and deployment environments. While these processes can be broadly defined, their implementation and implications are not bound by absolute characteristics. After all, there is no such thing as 100% testing, quality, security, or feedback. These aspects are always relative to specific objectives and contexts. Here are the challenges:

- **The challenges of defining continuous testing**:

  - **Varied testing needs**: The scope and method of software testing vary greatly, depending on the type of software, its intended use, user base, and the development methodology employed. For instance, testing for a safety-critical system such as aviation software differs vastly from testing a consumer mobile application.

  - **Evolving technologies**: As technology evolves, so do the testing methodologies. New paradigms such as AI and IoT bring new testing challenges that were not considered in traditional testing frameworks.

- **The challenges of defining continuous quality**:

  - **Subjective nature**: Quality is inherently subjective and can be viewed differently by different stakeholders. For a developer, it might mean code readability and maintainability, while for end users, it's about usability and performance.

  - **Context-dependent**: The quality standards for a rapidly developed prototype may not be the same as for a mature, customer-facing product. The context of development and deployment plays a crucial role in determining what constitutes quality.

- **The challenges of defining continuous security**:

  - **Changing threat landscape**: The landscape of cybersecurity threats is continually evolving. What is considered secure today may not be secure tomorrow, making it impossible to achieve absolute security.

  - **Risk management**: Security is often about managing risk rather than eliminating it. Different applications require different levels of security, based on their exposure to threats and the sensitivity of the data they handle.

- **Challenges of defining continuous feedback:**

  - **Diverse sources and interpretations**: Feedback can come from various sources (users, stakeholders, and automated systems) and can be interpreted in many ways. What is valuable feedback in one scenario might be irrelevant in another.

  - **Continuous adaptation**: Feedback mechanisms must adapt to the changing needs and expectations of users and the market. This means that the process of gathering and implementing feedback is never complete and always subject to change.

While processes of testing, quality, security, and feedback for continuous delivery and continuous operations can be defined, they do not possess absolute characteristics. They are highly context-dependent and must be aligned with specific objectives, technological environments, and user expectations. This inherent variability and the need for constant adaptation make it challenging to standardize these concepts across all software development and operation scenarios.

A definition of continuous testing, quality, security, and feedback

In the dynamic field of software engineering, particularly with continuous delivery (DevOps and DevSecOps) practices and continuous operation (SRE) practices, it's crucial to focus on outcomes rather than just process actions. Many existing definitions tend to concentrate excessively on procedural aspects, overlooking the importance of aligning with business outcomes. A more practical and useful approach involves defining strategies for continuous testing, continuous quality, continuous security, and continuous feedback in a way that emphasizes measurable business outcomes. These outcomes, particularly aligned with the **DevOps Research Association's (DORA's)** metrics, are critical in assessing the efficiency and success of software development practices. With these considerations in mind, the following definitions can be used in this document:

- **A continuous testing definition**: Continuous testing is a strategy designed to reduce lead times and failure rates in continuous delivery pipelines and continuous operations, through automated and iterative testing processes, aiming for decreased time from code commit to production deployment and reduced failures in production:

  - *Metrics*:

    - Time spent on testing tasks, from code commit to production deployment.
    - The percentage of defects that escape to production.

  - *Rationale*:

    - This definition integrates testing into every stage of development, delivery, and operations, ensuring early and consistent detection and resolution of issues, which is crucial for rapid and reliable software delivery and operations.

- **A continuous quality definition**: Continuous quality is a strategy to enhance user satisfaction and reduce production failure rates by integrating quality metrics throughout the development, delivery, and production processes, focusing on stable releases with fewer user issues:

  - *Metrics*:

    - The rate of releases approved for deployment

    - Customer-reported issues per release

    - Availability level objectives (SLOs)

  - *Rationale*:

    - By prioritizing quality at every phase of development and operations, this strategy ensures the delivery of stable and reliable software, meeting user expectations and business needs.

- **A continuous security definition**: Continuous security is a strategy that integrates security measures into continuous development, delivery, and operations to reduce the frequency and impact of security events, measured by security events and security event resolution times:

  - *Metrics*:

    - The number of (pre- and post-release) security events

    - The mean time to detect, respond, and resolve security events

  - *Rationale*:

    - This strategy underscores proactive security practices, embedding security considerations into the entire software life cycle, essential for maintaining software integrity and trust.

- **A continuous feedback definition**: Continuous Feedback is a strategy that utilizes stakeholder and user feedback to accelerate release frequency and improve recovery times, measured by the implementation speed of feedback and its impact on system reliability:

  - *Metrics*:

    - Time to implement feedback (source to resolver)

    - The rate of releases approved for deployment

    - Customer-reported issues per release

    - Availability level objectives (SLOs)

- *Rationale*:

  - A systematic collection and implementation of feedback ensure that the software continually evolves in response to user needs and market changes, driving continuous improvement.

Figure 1.4 provides practical definitions for continuous testing, quality, security, and feedback, as used in this document.

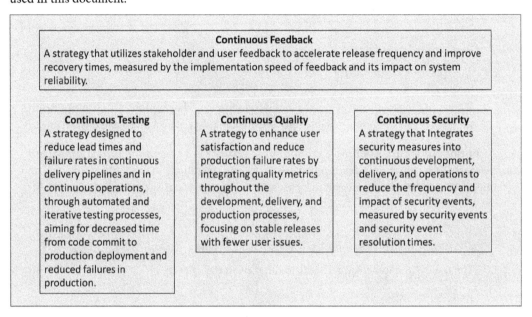

Figure 1.4 – Continuous testing, quality, security and feedback definitions

Adopting these strategically focused definitions for continuous testing, quality, security, and feedback allows organizations to align their continuous development, delivery, and operations practices with measurable business outcomes. This approach not only provides a clear direction for continuous improvement but also ensures that the methodologies are evaluated and adjusted, based on their impact on key performance indicators. In the evolving landscape of software development, such outcome-driven strategies are indispensable to achieve efficiency, reliability, and success in digital transformations.

The guiding principles and pillars of continuous testing

This section describes the guiding principles and pillars of practice that are important to support an effective continuous testing strategy. They are essential for ensuring that continuous testing effectively decreases the time from code commit to production deployment and reduces failures in production.

Figure 1.5 illustrates the pillars of continuous testing.

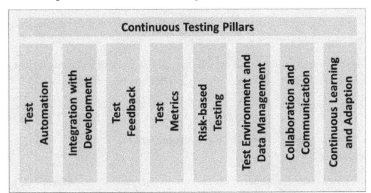

Figure 1.5 – The continuous testing pillars

Let's look at them in detail:

- **Test automation**:

  - *Principle*: Automation is key to achieving the speed and consistency required for continuous testing. Automated tests can be run frequently and consistently, ensuring rapid feedback on the health of the software.

  - *Pillar*: Develop and maintain a suite of automated tests that cover a wide range of aspects, including unit, integration, regression, performance, security, system, and user acceptance testing.

- **Integration with a development life cycle**:

  - *Principle*: Testing should be integrated into the development process from the very beginning, not tacked on at the end.

  - *Pillar*: Implement a shift-left approach, where testing starts early in the development cycle. This includes practices such as **Test-Driven Development** (**TDD**) and **Behavior-Driven Development** (**BDD**).

- **Test feedback**:

  - *Principle*: Continuous feedback from testing is vital for the timely identification and resolution of issues.

  - *Pillar*: Establish mechanisms for real-time reporting and analysis of test results, ensuring immediate action can be taken when issues are identified. Actions such as bug and vulnerability issue reports can be automated.

- **Testing metrics**:

    - *Principle*: Metrics and measurement are essential to understand the effectiveness of testing efforts and to guide continuous improvement.

    - *Pillar*: Use a comprehensive set of quality metrics, such as defect density, test coverage, and mean time to resolution, to track and improve the testing process.

- **Risk-based testing**:

    - *Principle*: Focus testing efforts on the most critical aspects of the application, based on risk assessment.

    - *Pillar*: Prioritize testing resources on areas with the highest risk or impact, such as critical functionality, performance bottlenecks, and security vulnerabilities.

- **Test environment and test data management**:

    - *Principle*: Reliable and consistent test environments and data are necessary for accurate testing.

    - *Pillar*: Ensure the availability of stable, scalable, and production-like test environments, along with appropriate test data management strategies.

- **Collaboration and communication**:

    - *Principle*: Effective collaboration and communication among developers, testers, and operations teams are vital for the success of continuous testing.

    - *Pillar*: Foster a culture of collaboration, where teams work together closely and share responsibility for quality.

- **Continuous learning and adaptation**:

    - *Principle*: Continuous testing is an evolving practice that should adapt to changing technologies and project requirements.

    - *Pillar*: Regularly review and adapt testing strategies, tools, and processes to meet the evolving needs of the software and the business.

These guiding principles and pillars of practice form the foundation of a robust continuous testing strategy. They help ensure that testing is efficient, effective, and aligned with the overall goals of reducing lead times, minimizing failures in production, and ultimately delivering high-quality software promptly.

The guiding principles and pillars of continuous quality

This section describes the guiding principles and pillars of practice that are important to support an effective continuous quality strategy.

Figure 1.6 illustrates the pillars of continuous quality.

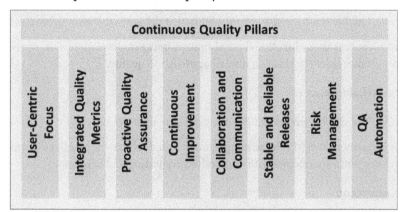

Figure 1.6 – The continuous quality pillars

Let's have a look at them:

- **User-centric focus**:

  - *Principle*: User satisfaction is a key indicator of quality.

  - *Pillar*: Regularly gather and analyze user feedback, usability testing results, customer satisfaction metrics, and results from satisfaction surveys to guide quality improvements.

- **Integrated quality metrics**:

  - *Principle*: Quality should be measurable and integrated into every stage of the software life cycle.

  - *Pillar*: Implement and continuously refine a set of quality metrics (such as defect rates, uptime, and performance benchmarks) across the development, delivery, and production phases.

- **Proactive quality assurance**:

  - *Principle*: Quality is not just about fixing defects; it's about preventing them.

  - *Pillar*: Employ proactive quality assurance practices, such as static code analysis, design reviews, and architectural evaluations, to identify and address potential issues early in the life cycle.

- **Continuous improvement**:

  - *Principle*: Quality is an evolving target that requires continuous improvement.

  - *Pillar*: Foster a culture of continuous improvement with regular retrospectives and reviews of processes, tools, and practices to identify areas for enhancement.

- **Collaboration and communication**:

  - *Principle*: Quality is a collective responsibility that demands collaboration across teams.

  - *Pillar*: Encourage cross-functional collaboration between developers, QA, operations, and business stakeholders to ensure a unified approach to quality.

- **Stable and reliable releases**:

  - *Principle*: The stability and reliability of releases are paramount.

  - *Pillar*: Implement robust release management and deployment practices to ensure stable and reliable software releases with comprehensive testing and validation.

- **Risk management**:

  - *Principle*: Identifying and managing risk is crucial to maintaining quality.

  - *Pillar*: Conduct regular risk assessments and prioritize efforts based on the potential impact on user satisfaction and system stability.

- **Quality Assurance (QA) automation**:

  - *Principle*: Automation is essential for scaling quality assurance practices.

  - *Pillar*: Utilize automated testing and quality assurance tools to increase coverage and efficiency, while freeing up resources to focus on complex quality challenges.

These guiding principles and pillars of practice define a comprehensive approach to continuous quality. By focusing on integrating quality metrics, emphasizing user satisfaction, promoting proactive quality assurance, and fostering continuous improvement, organizations can effectively enhance the overall quality of their software products, leading to fewer production failures and higher user satisfaction.

The guiding principles and pillars of continuous security

This section describes the guiding principles and pillars of practice that are important to support an effective continuous security strategy.

Figure 1.7 illustrates the pillars of continuous security.

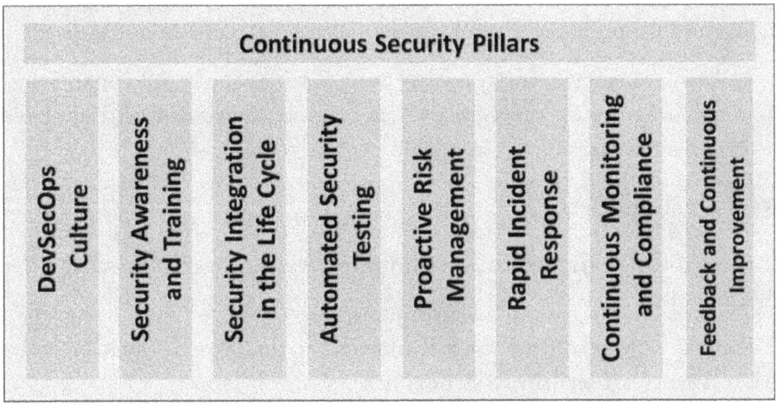

Figure 1.7 – The continuous security pillars

Let's look at them in brief:

- **DevSecOps culture**:

  - *Principle*: Collaboration between development, security, and operations enhances security outcomes.

  - *Pillar*: Promote a DevSecOps culture where security is a shared responsibility, integrated into the DevOps practices, encouraging collaboration and communication across teams.

- **Security awareness and training**:

  - *Principle*: Security is a shared responsibility and requires awareness at all levels.

  - *Pillar*: Provide regular security training and awareness programs for all team members to foster a security-conscious culture. For example, security training on topics such as **Open Worldwide Application Security Project (OWASP)** training, secure coding, and API security can be important.

- **Security integration in the life cycle**:

  - *Principle*: Security is an integral part of the entire software life cycle, not an isolated stage.

  - *Pillar*: Embed security practices and tools into the development, delivery, and operational processes, ensuring that security considerations are addressed from inception through to deployment and maintenance.

- **Automated security testing**:
  - *Principle*: Automation is key to maintaining continuous security vigilance.
  - *Pillar*: Utilize automated security testing tools (such as static and dynamic analysis tools and vulnerability scanners) to regularly scan and identify security threats at every stage of the development process.

- **Proactive risk management**:
  - *Principle*: Proactive identification and mitigation of risks are more effective than reactive measures.
  - *Pillar*: Conduct regular security risk assessments and threat modeling to proactively identify and address potential security vulnerabilities.

- **Rapid incident response**:
  - *Principle*: Quick and effective response to security incidents minimizes their impact.
  - *Pillar*: Establish a well-defined incident response plan that includes procedures for rapid detection, investigation, and remediation of security events.

- **Continuous monitoring and compliance**:
  - *Principle*: Ongoing monitoring and adherence to compliance standards are critical to maintain security.
  - *Pillar*: Implement continuous monitoring solutions to detect and alert about suspicious activities, along with regular compliance checks to ensure adherence to relevant security standards and regulations.

- **Feedback and continuous improvement**:
  - *Principle*: Feedback is essential for the evolution and improvement of security practices.
  - *Pillar*: Implement feedback mechanisms to learn from security events, and continuously improve security measures based on lessons learned and evolving threats.

These guiding principles and pillars establish a robust framework for continuous security. They ensure that security is a continuous, integrated process, emphasizing proactive risk management, rapid incident response, and ongoing monitoring, while fostering a culture of security awareness and collaboration. By adhering to these principles, organizations can effectively reduce the frequency and impact of security events, thereby enhancing their overall security posture.

The guiding principles and pillars of continuous feedback

This section describes the guiding principles, and pillars of practice that are important for an effective continuous feedback strategy.

Figure 1.8 illustrates the pillars of Continuous Feedback.

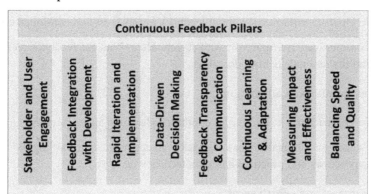

Figure 1.8 – The continuous feedback pillars

Let's discuss these pillars next:

- **Stakeholder and user engagement**:

  - *Principle*: Active engagement with stakeholders and users is essential for relevant and actionable feedback.

  - *Pillar*: Establish regular channels to gather feedback from all stakeholders, including customers, end users, team members, and business partners.

- **Feedback integration with development**:

  - *Principle*: Feedback should be integrated seamlessly into the development process.

  - *Pillar*: Develop mechanisms to quickly integrate feedback into the development pipeline, ensuring that it directly informs development priorities and decisions.

- **Rapid iteration and implementation**:

  - *Principle*: The value of feedback is maximized when it is implemented rapidly and effectively.

  - *Pillar*: Focus on shortening the cycle time from receiving feedback to implementing changes, enabling faster iterations and improvements.

- **Data-driven decision making**:

  - *Principle*: Decisions should be based on data derived from feedback, not just intuition or assumptions.

  - *Pillar*: Utilize tools and techniques to analyze feedback quantitatively and qualitatively, ensuring that decisions are informed by actual user and stakeholder insights.

- **Feedback transparency and communication**:

  - *Principle*: Open communication about feedback fosters trust and continued engagement.

  - *Pillar*: Communicate transparently with stakeholders about the feedback received, actions taken, and the rationale behind decisions.

- **Continuous learning and adaptation**:

  - *Principle*: Feedback is a key driver for continuous learning and adaptation.

  - *Pillar*: Encourage a culture that views feedback as an opportunity for learning and improvement, adapting processes and practices based on feedback insights.

- **Measuring impact and effectiveness**:

  - *Principle*: The effectiveness of feedback implementation should be continually measured.

  - *Pillar*: Track and evaluate the impact of feedback on release frequency, recovery times, and system reliability to measure the effectiveness of feedback implementation.

- **Balancing speed and quality**:

  - *Principle*: While rapid implementation of feedback is important, maintaining quality is equally crucial.

  - *Pillar*: Ensure that feedback is implemented in a way that balances speed with the need to maintain or enhance the quality and reliability of a system.

These guiding principles and pillars form a comprehensive framework for continuous feedback, emphasizing the importance of stakeholder and user engagement, rapid integration of feedback into development, and data-driven decision making. By adhering to these principles, organizations can effectively use feedback to drive faster releases, improve recovery times, and enhance overall system reliability, thereby aligning closely with the goals of modern software development methodologies.

Summary

In the rapidly evolving landscape of DevOps, DevSecOps, and SRE, strategies for continuous testing, quality, security, and feedback have emerged as pivotal elements in steering digital transformations toward successful continuous development, delivery, and operations. This chapter delved into the heart of these strategies, offering practical definitions and the guiding principles that underpin them.

The journey began with *Introducing Continuous Testing, Quality, Security, and Feedback*, setting the stage for a comprehensive exploration. This section laid the groundwork, illuminating why these concepts are indispensable in modern software development. It's an invitation to view software development, delivery, and operations through a lens that prioritizes continuous improvement and adaptation. The following section, *Defining Continuous Testing, Quality, Security, and Feedback*, provided clear, outcome-focused definitions of each concept. This clarity is crucial, as it serves as a beacon for professionals navigating the complexities of implementing these strategies, which is essential for digital transformation.

The heart of the chapter lies in the detailed exposition of the guiding principles and pillars for each concept. *Guiding Principles and Pillars of Continuous Testing* explained that integrating testing into every stage of the software development life cycle ensures that quality and functionality are not afterthoughts but ingrained in the process. The section on continuous quality emphasized a proactive approach to maintaining high standards, ensuring that a product not only meets but exceeds user expectations. When it comes to continuous security, the chapter underscored the need for an integrated, vigilant approach to protect against evolving threats.

In the segment dedicated to continuous feedback, the chapter highlighted the significance of stakeholder and user input in shaping and refining software products. This feedback loop is depicted as a dynamic, integral component of the development process, driving improvements and fostering user satisfaction. Finally, the chapter equipped you with valuable skills – understanding the essence of continuous testing, quality, security, and feedback and learning to implement their guiding principles effectively. This knowledge is not just theoretical; it's a toolkit to thrive in the modern software development arena.

In summary, this chapter is a practical guide and a catalyst for transformation. It encourages you to embrace these continuous strategies, ensuring digital transformations are resilient, user-centric, and secure. Whether you're a seasoned professional or just starting, this chapter provided valuable insights and skills that will elevate your approach to continuous software development, delivery, and operations.

The next chapter explains why continuous testing, quality, security, and feedback are essential for DevOps, DevSecOps, and SRE.

The Importance of Continuous Testing, Quality, Security, and Feedback

This chapter explains why **Continuous Testing**, **Quality**, **Security**, and **Feedback** strategies are important for **DevOps**, **DevSecOps**, and **Site Reliability Engineering (SRE)**. To understand why, it is necessary to understand how the principles and pillars of DevOps, DevSecOps, and SRE depend on the principles and pillars of Continuous Testing, Quality, Security, and Feedback. This is simplified, to some extent, because the pillars for the **continuous delivery (CD)** strategies such as DevOps and DevSecOps are similar and therefore can be grouped together for the purpose of comparison. The principles and pillars for continuous operations equate to SRE so these are compared in a separate section.

The chapter is organized into three sections:

- Why continuous strategies are important for DevOps and DevSecOps
- Why continuous strategies are important for SRE
- Consequences of implementing DevOps, DevSecOps, and SRE without properly implementing continuous practices

Let's get started!

Why continuous strategies are important for DevOps and DevSecOps

This section introduces principles and pillars of practices for CD strategies DevOps and DevSecOps. The section explains why those principles and pillars of practices depend on the principles and pillars of practices for continuous testing, quality, security, and feedback strategies.

Principles and pillars of DevOps, and DevSecOps

DevOps and DevSecOps are related strategies that both facilitate CD, though they emphasize different aspects of the software development and delivery process.

The nine pillars of DevOps practices

DevOps primarily focuses on improving collaboration between software **development** (**Dev**) and **operations** (**Ops**) teams. Its goal is to streamline the development pipeline, implement automation, and enable faster and more efficient delivery of software.

As defined by Marc Hornbeek in *Engineering DevOps* (self-published with Bookbaby in 2019), the nine pillars of DevOps represent fundamental principles and practices that are essential for the successful implementation of DevOps in an organization. Each pillar plays a unique role in supporting and enhancing the DevOps philosophy, which focuses on improving collaboration, automation, and the overall efficiency of software development and delivery. Here's an explanation for each of these pillars:

- **Leadership**: Effective leadership is crucial for driving the cultural and organizational change required for DevOps. Leaders must advocate for DevOps principles, facilitate collaboration across departments, and ensure that teams are equipped with the necessary tools and training. They play a key role in breaking down silos and fostering an environment that supports continuous improvement and innovation.

- **Collaborative culture**: DevOps emphasizes a culture of collaboration and communication between software developers, IT operations, and other stakeholders in the software delivery process. A collaborative culture involves shared responsibilities, transparency, and open communication, which are essential for rapid and efficient development, testing, and deployment of software.

- **Design for DevOps**: This involves designing software and systems in a way that supports DevOps practices. It includes considerations such as modularity, scalability, and maintainability. The design should facilitate easy updates, quick deployments, and efficient operation, aligning with the principles of automation and CD.

- **Continuous integration (CI)**: CI is the practice of frequently integrating code changes into a shared repository. Each integration is verified by an automated build and testing process to detect integration errors as quickly as possible, thereby reducing integration problems and improving software quality.

- **Continuous testing**: Continuous testing involves the execution of automated tests as part of the software delivery pipeline. This ensures that software quality is maintained throughout the development process, allowing teams to identify and address issues early, and enabling faster and more reliable software releases.

- **Elastic infrastructure**: Elastic infrastructure refers to the ability to dynamically scale resources up or down based on demand. In a DevOps context, this means using virtualized and cloud-based services and **infrastructure-as-a-service** technologies, which allow for the flexible allocation of computing resources, supporting the rapid and efficient deployment and operation of applications.

- **Continuous security**: Continuous security, or DevSecOps, integrates security practices into every step of the software development and deployment process. It involves continuous monitoring for security threats, regular vulnerability assessments, and the integration of security controls into the CI/CD pipeline.

- **Continuous delivery (CD)**: CD is a software engineering approach where teams produce software in short cycles, ensuring that it can be reliably released at any time. It aims to build, test, and release software with greater speed and frequency, reducing the cost, time, and risk of delivering changes.

- **Continuous monitoring**: Continuous monitoring involves the ongoing collection, processing, and analysis of data (performance metrics, logs, etc.) from software applications and infrastructure. This practice helps in proactively identifying and addressing issues, understanding system performance, and ensuring that the system operates reliably and efficiently.

Together, these nine pillars form a comprehensive framework for implementing and optimizing DevOps practices. They encourage a holistic approach to software development and delivery, focusing on collaboration, automation, continuous improvement, and a high degree of operational efficiency. By integrating these practices, DevOps aims to shorten the development cycle, enhance software quality, and increase the frequency of releases, thereby responding more quickly to market demands and user needs.

The nine pillars of DevSecOps practices

DevSecOps extends the DevOps model by integrating security practices into all DevOps pillars. This strategy emphasizes that security should not be an afterthought but a key component of the development process. DevSecOps seeks to embed automated security checks and balances within the CI/CD pipeline, ensuring that security considerations are addressed continuously and do not slow down the delivery process.

As defined by Marc Hornbeek in *Engineering DevOps*, the nine pillars of DevOps also apply to DevSecOps. While the pillars are the same, the specific practices under each of the pillars complement each other. The nine pillars of DevSecOps ensure that security is not a standalone element but is seamlessly embedded into every stage of the software development and delivery pipeline. Here are explanations for each of these pillars of DevSecOps:

- **Leadership**: Strong leadership is crucial in driving the cultural shift necessary for the successful integration of security into DevOps practices. Leaders must champion security as a core value, ensuring it is prioritized at every level of the organization and that teams have the resources and support they need to implement DevSecOps effectively.

- **Collaborative culture**: A culture of collaboration is essential in DevSecOps, where development, operations, and security teams work closely together. This collaboration ensures a shared understanding and responsibility for security, enabling faster and more effective identification and resolution of security issues.

- **Design for DevOps**: This involves designing systems and applications with both DevOps and security considerations in mind. It includes building with modularity, scalability, and security from the start, ensuring that systems are both agile and secure.

- **Continuous integration (CI):** In DevSecOps, CI includes integrating code changes frequently and ensuring each change is automatically tested for security issues. This approach helps in identifying and addressing security vulnerabilities early in the development process.

- **Continuous testing**: Continuous testing in DevSecOps extends beyond functionality and performance to include regular, automated security testing. This ensures that security is continuously validated at every stage of the software development life cycle.

- **Elastic infrastructure**: Elastic infrastructure in DevSecOps emphasizes the ability to rapidly scale resources in response to varying demands while maintaining security. It involves using cloud services and virtualization, with version and flexible configuration controls that support both operational flexibility and security requirements.

- **Continuous security**: Continuous security is at the heart of DevSecOps. It involves integrating security practices into every phase of the software development and deployment process, from initial design through to operation, ensuring ongoing monitoring and compliance.

- **Continuous delivery (CD)**: In the context of DevSecOps, CD entails a process where code changes are automatically built, tested (including security testing), and prepared for release to production, ensuring that software can be deployed at any time with high security.

- **Continuous monitoring**: This involves constantly monitoring the deployed software and underlying infrastructure for security threats. Continuous monitoring enables rapid detection and response to security incidents, vulnerabilities, and anomalies.

Each of these pillars plays a vital role in embedding security into the DevOps pipeline, ensuring that security considerations are an integral and ongoing part of the development and deployment process. Together, they form a comprehensive framework for implementing DevSecOps effectively, aligning security with the fast-paced and agile nature of modern software development.

Alignment of DevOps and DevSecOps

Both DevOps and DevSecOps share the common goal of facilitating CD but approach it with different emphases: DevOps prioritizes efficiency and speed, while DevSecOps prioritizes security within that fast-paced environment.

While it is feasible (and there are unfortunately many industry examples of this) to implement DevOps without DevSecOps, there are concerns with this:

- **Security as an afterthought**: In a pure DevOps approach, the focus on speed and efficiency might lead to security being overlooked or tacked on at the end of the development process. This can result in vulnerabilities being introduced into the software, potentially leading to security breaches and deploys when vulnerabilities found late in the pipeline need to be fixed, integrated, and verified.

- **Reactive rather than proactive security**: Without integrating security into the CD pipeline, organizations may find themselves reacting to security incidents post-deployment, which can be costly and damaging to the organization's reputation.

- **Compliance risks**: A lack of emphasis on security may lead to non-compliance with industry regulations and standards, particularly in sectors where data protection and privacy are crucial.

- **Increased vulnerability to cyber threats**: CD without continuous security can make software products more vulnerable to emerging cyber threats, as security measures may not keep pace with the rapid deployment of updates and new features.

- **A fragmented approach to risk management**: Separating development and security efforts can lead to a fragmented approach to risk management, where security considerations are not adequately integrated into the decision-making process during development.

In summary, while DevOps and DevSecOps both aim to facilitate Continuous Delivery, the integration of security from the outset – as emphasized in DevSecOps – is crucial to ensure that the speed and efficiency gains of DevOps do not compromise the security and integrity of the software being delivered. Thus, a holistic approach that combines the principles of both DevOps and DevSecOps with a common framework of nine pillars is essential for a balanced, efficient, and secure software delivery pipeline.

Figure 2.1 from *Engineering DevOps* illustrates the nine pillars of DevOps and DevSecOps. As indicated in the figure, the nine pillars model shows how all the pillars share a common foundation of orchestration, automation, and governance, and a common "roof" of a CI/CD pipeline, application release automation, and value stream management.

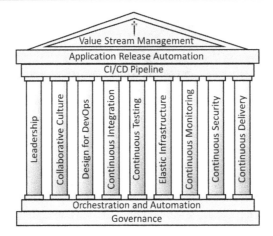

Figure 2.1 – The nine pillars of DevOps and DevSecOps

The next subsection explains how the nine pillars of DevOps and DevSecOps depend on continuous testing, quality, security, and feedback.

DevOps and DevSecOps dependencies on continuous testing, quality, security, and feedback

The nine pillars of DevOps and DevSecOps intersect with and are supported by the pillars of continuous testing, continuous quality, continuous security, and continuous feedback in various ways. Each set of pillars complements and reinforces the others, creating a comprehensive approach to software development and operations. Here's how they are interconnected:

- **Leadership**: This supports all pillars by providing strategic direction, resources, and advocacy for testing, quality, security, and feedback initiatives.

- **Collaborative culture**: This aligns with collaboration and communication in continuous testing, quality, and security, as well as stakeholder and user engagement in continuous feedback, fostering a shared responsibility for these aspects.

- **Design for DevOps**: This relies on proactive quality assurance, risk-based testing, and security integration in the life cycle, ensuring systems are designed to facilitate these practices.

- **Continuous integration (CI)**: This is directly supported by test automation, integration with development, and automated security testing, enabling frequent and reliable integration of changes.

- **Continuous testing**: This embodies the principles of the pillars of continuous testing, ensuring that testing is automated, integrated, and continuously improved.

- **Elastic infrastructure**: This is supported by test environment and test data management in continuous testing, as well as continuous monitoring and compliance in continuous security, ensuring infrastructure can adapt while maintaining quality and security standards.

- **Continuous security**: This encompasses the pillars of continuous security, integrating security practices at every step of the DevOps pipeline.

- **Continuous delivery (CD)**: This relies on stable and reliable releases from continuous quality, as well as feedback integration with development and rapid incident response from continuous security, to ensure that software can be released quickly and securely.

- **Continuous monitoring**: This is supported by continuous monitoring and compliance in continuous security and measuring impact and effectiveness in continuous feedback, providing insights into the performance, quality, and security of the system.

These interdependencies show that the nine pillars of DevOps and DevSecOps are not isolated elements but are intrinsically connected to the foundational aspects of continuous testing, quality, security, and feedback.

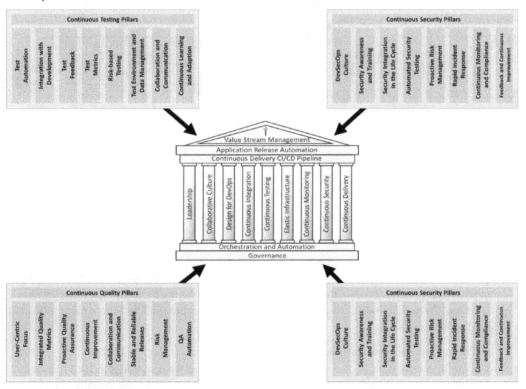

Figure 2.2 – DevOps and DevSecOps dependencies on continuous testing, quality, security, and feedback

Each set of pillars enhances and strengthens the others, creating a holistic and integrated approach to software development and delivery. This synergy is crucial for achieving the goals of DevOps and DevSecOps: fast, reliable, secure, and high-quality software delivery that meets user needs and business objectives.

Why continuous strategies are important for SRE

This section introduces principles and pillars of practice for continuous operations SRE strategies. The section explains why those principles and pillars of practices depend on the principles and pillars of practices for continuous testing, quality, security, and feedback strategies.

Principles and pillars of SRE

SRE is a discipline that incorporates aspects of software engineering and applies them to infrastructure and operations problems. The goal is to create scalable and highly reliable software systems.

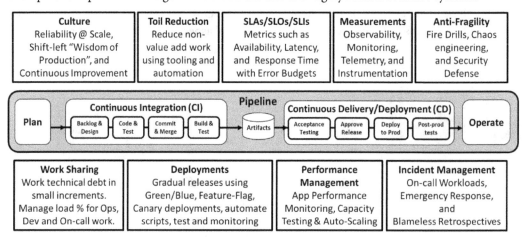

Figure 2.3 – Principles and pillars of SRE

Here's an explanation of the nine pillars of SRE as illustrated in *Figure 2.3*:

- **Culture – shift left wisdom of production**: This pillar emphasizes the importance of incorporating operational knowledge early in the software development life cycle. *Shifting left* means considering reliability, scalability, and operational aspects from the beginning of the development process, not just at the end or after deployment. It fosters a culture where production wisdom informs development decisions.

- **Toil reduction and automation**: Toil refers to the repetitive, manual, and non-strategic work that can be automated. Reducing toil through automation frees up SREs to focus on more impactful work that improves system reliability and efficiency. This pillar underscores the importance of identifying and automating routine tasks.

- **SLAs / SLOs / SLIs and error budgets**: Service-level agreements (SLAs), **service-level objectives (SLOs)**, and **service-level indicators (SLIs)** are critical for measuring the reliability of services. SLAs are commitments to customers, SLOs are goals for service levels, and SLIs are the metrics used to measure these levels. Error budgets define the acceptable threshold for errors and help balance the need for reliability with the pace of innovation.

- **Measurements and observability**: This pillar focuses on the ability to measure and observe the internal state of systems and infrastructures. Effective observability is crucial for understanding system performance and behavior, which in turn informs decisions about reliability and efficiency.

- **Anti-fragility, fire drills, chaos engineering, and security defense**: Anti-fragility goes beyond resilience by gaining from stressors and challenges. Practices such as fire drills and chaos engineering (intentionally injecting failure to test systems) are used to make systems more robust. This pillar also emphasizes the importance of proactive security defense measures.

- **Work sharing and incremental technical debt**: This pillar advocates for collaborative work sharing across development and operations teams to spread knowledge and responsibility. It also emphasizes the importance of managing technical debt incrementally, preventing it from accumulating to unmanageable levels.

- **Deployments using blue-green, A/B, and Canary**: These deployment strategies are used to reduce the risk associated with releasing new versions of software. Blue-green, A/B testing, and Canary releases allow for controlled exposure of new changes to users, enabling testing in production environments and quick rollback if necessary.

- **Performance management of apps and infrastructure**: This involves monitoring and optimizing the performance of applications and underlying infrastructure. It's about ensuring that both the software and the hardware it runs on are optimized for efficiency, scalability, and reliability. This pillar includes capacity planning in which SREs determine whether the systems will withstand peak loads and implement mitigation strategies to cope with surges of demand.

- **Incident management, on-call, emergencies, and retrospectives**: Effective incident management and having a structured on-call response for emergencies are crucial in SRE. This pillar also emphasizes the importance of conducting retrospectives after incidents to learn and improve. It's about creating a process for responding to, learning from, and preventing future incidents.

These pillars of SRE provide a framework for building and maintaining reliable systems. They emphasize proactive and preventive measures, continuous improvement, and a balanced approach to innovation and stability.

SRE dependencies on continuous testing, quality, security, and feedback

The nine pillars of SRE are closely intertwined with the principles and pillars of continuous testing, continuous quality, continuous security, and continuous feedback.

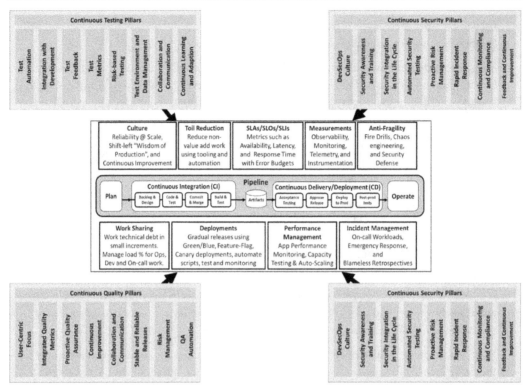

Figure 2.4 – SRE dependencies on continuous testing, quality, security, and feedback

Figure 2.4 and the following explanation show how they are interconnected:

- **Culture – shift left wisdom of production**:

  - **Continuous testing**: Integration with development and continuous learning and adaptation are essential for shifting left.

  - **Continuous quality**: Proactive quality assurance ensures that quality considerations are made early.

  - **Continuous security**: Security integration in the life cycle aligns with shifting security considerations left in the development process.

- **Toil reduction and automation**:

  - **Continuous testing**: Test automation is key in reducing manual testing efforts.

  - **Continuous quality**: Quality assurance automation helps in automating quality checks.

  - **Continuous security**: Automated security testing reduces manual security checks.

- **SLAs / SLOs / SLIs and error budgets**:

  - **Continuous feedback**: Data-driven decision making and measuring impact and effectiveness align with defining and monitoring SLAs, SLOs, and SLIs.

  - **Continuous quality**: Integrated quality metrics support the establishment of effective SLOs.

- **Measurements and observability**:

  - **Continuous testing**: Test metrics provide critical data for observability.

  - **Continuous quality**: Continuous improvement relies on effective measurements.

  - **Continuous security**: Continuous monitoring and compliance are integral to observability.

- **Anti-fragility, fire drills, chaos engineering, and security defense**:

  - **Continuous testing**: Risk-based testing aligns with preparing for worst-case scenarios.

  - **Continuous security**: Proactive risk management and rapid incident response are crucial here.

  - **Continuous feedback**: Continuous learning and adaptation are essential for evolving from these activities.

- **Work sharing and incremental technical debt**:

  - **Continuous testing**: Collaboration and communication facilitate work sharing.

  - **Continuous quality**: Managing technical debt is part of continuous improvement and risk management.

- **Deployments using blue-green, A/B, and Canary**:

  - **Continuous testing**: Test environment and test data management support these deployment strategies.

  - **Continuous quality**: Stable and reliable releases are crucial for successful blue-green, A/B, and Canary deployments.

- **Performance management of apps and infrastructure**:

  - **Continuous testing**: Test automation and test metrics contribute to performance management.

  - **Continuous quality**: User-centric focus and integrated quality metrics are key in ensuring that performance aligns with user expectations.

- **Incidents management, on-call, emergencies, and retrospectives**:

  - **Continuous feedback**: Feedback transparency and communication, as well as continuous learning and adaptation, are critical for effective incident management and learning from retrospectives.

  - **Continuous security**: Rapid incident response and continuous monitoring and compliance are foundational to this pillar.

These interdependencies illustrate that SRE's pillars are not standalone elements but are deeply connected to continuous testing, quality, security, and feedback. Each set of pillars complements the others, creating a comprehensive approach to reliable, secure, and efficient software development and operations.

Consequences of implementing DevOps, DevSecOps, and SRE without properly implementing continuous practices

Implementing DevOps, DevSecOps, and SRE without properly integrating the pillars of continuous testing, quality, security, and feedback can lead to a range of negative consequences. These methodologies are designed to work synergistically with these continuous practices, and neglecting any aspect can undermine their effectiveness. Here are examples of potential consequences for each methodology:

- **DevOps without proper continuous practices**:

  - **Inadequate testing**: If continuous testing is not properly implemented in DevOps, it can lead to frequent production failures due to bugs and performance issues, undermining the goal of rapid and reliable software delivery.

  - **Poor quality**: Neglecting continuous quality can result in software that does not meet user expectations or has poor usability, leading to diminished customer satisfaction and potential loss of trust.

  - **Security vulnerabilities**: Without continuous security, DevOps might accelerate the deployment of insecure software, increasing the risk of data breaches and cyberattacks.

  - **Misalignment with user needs**: Lack of continuous feedback can lead to a disconnect between what is developed and what users actually need, resulting in products that fail to address market demands effectively.

- **DevSecOps without proper continuous practices**:

  - **Security gaps**: Ineffective implementation of continuous security practices in DevSecOps can leave vulnerabilities unchecked, exposing the organization to security risks and compliance issues.

- **Delayed response to security incidents**: Without continuous feedback and monitoring, there can be a slower response to security incidents, exacerbating the impact of breaches.

- **Quality and testing overlooked**: If continuous testing and quality are not integrated, even a security-focused pipeline can deploy software with functional defects or performance issues, affecting user satisfaction and system reliability.

- **SRE without proper continuous practices**:

  - **System instability**: Failing to integrate continuous testing and quality in SRE can result in systems that are not robust or scalable, leading to frequent outages or performance degradation.

  - **Inefficient operations**: Without continuous feedback and the reduction of toil through automation, SRE teams might spend excessive time on operational firefighting rather than on strategic improvements, leading to burnout and inefficiency.

  - **Inadequate security measures**: Neglecting continuous security within SRE practices can make systems susceptible to security threats, compromising both system reliability and user trust.

  - **Poor incident management**: Lack of continuous feedback mechanisms can hinder effective incident management, resulting in prolonged system downtimes and slower recovery from failures.

In summary, the absence of robust continuous testing, quality, security, and feedback practices can significantly diminish the effectiveness of DevOps, DevSecOps, and SRE implementations. It can lead to the deployment of unreliable, insecure, and low-quality software, increased operational challenges, and a failure to meet user and business needs effectively. Therefore, it's essential to fully integrate these continuous practices to realize the full benefits of DevOps, DevSecOps, and SRE methodologies.

Summary

This chapter has elucidated the intrinsic connection between DevOps, DevSecOps, SRE, and the continuous practices of testing, quality, security, and feedback, demonstrating their interdependence in achieving efficient, secure, and reliable software delivery.

DevOps, with its emphasis on bridging development and operations through collaboration and automation, inherently relies on the CI of testing and quality assurance to ensure fast and efficient delivery of software. The integration of continuous testing and quality practices within DevOps workflows enables organizations to maintain high standards of software performance and reliability, even as they accelerate their development cycles.

DevSecOps extends this model by weaving security into the fabric of DevOps practices. The principles of continuous security, such as proactive risk management and automated security testing, are essential for DevSecOps, ensuring that security measures are not compromised in the pursuit of speed and efficiency. Continuous security provides a framework for embedding security at every stage of the software development life cycle, making it a fundamental aspect of DevOps practices and not an afterthought.

SRE, focusing on the reliability and scalability of software systems, draws extensively on continuous practices to achieve its goals. continuous testing and quality are pivotal in setting and maintaining high standards for system performance and stability. Moreover, SRE's commitment to minimizing toil through automation and its emphasis on performance management are complemented by continuous feedback practices. These practices ensure that SRE is not only about maintaining the status quo but is also continuously learning and adapting to new challenges and user needs.

The chapter has highlighted that while DevOps, DevSecOps, and SRE have distinct focuses and methodologies, their success is fundamentally dependent on the integration of continuous practices. Continuous testing, quality, security, and feedback are not standalone strategies, but are interwoven into the life cycle of software development and operations in these approaches. They provide the tools, processes, and cultural mindset necessary for organizations to develop and maintain software that is not only fast and efficient but also robust, secure, and aligned with user expectations.

In conclusion, this chapter makes it clear that the future of software development and operations lies in the harmonious integration of DevOps, DevSecOps, and SRE with continuous testing, quality, security, and feedback. These strategies and practices, when implemented together, form a powerful synergy that drives innovation, enhances operational efficiency, and ensures the delivery of high-quality, reliable, and secure software products. They represent a comprehensive approach to modern software engineering, one that is adaptable, resilient, and user centric.

The next chapter will explain my lifetime of personal experiences and pitfalls with continuous testing, quality, security and feedback strategies.

Experiences and Pitfalls with Continuous Testing, Quality, Security, and Feedback

Looking back over my career of more than 48 years, I think I was destined to focus on what is now called **continuous testing**, **DevOps**, **DevSecOps**, and **SRE**, but these labels did not exist until more recently. Throughout that time, I primarily focused on designing, developing, and implementing testing, quality, security, and feedback strategies, and automated platforms and tools for software development, delivery, and operations. My career developed, coincident with the evolution from software waterfall to continuous delivery and operations methodologies. I was able to use and contribute to the development of these strategies, platforms, and tools, which ultimately resulted in being honored by the IEEE as the Outstanding Engineer for IEEE Region 6, 2016 for my lifetime of work on automation.

This chapter explains, by way of examples from my lifetime of experiences, use cases, lessons learned, and pitfalls to avoid, including strategies to avoid these pitfalls.

The first sections of the chapter are organized in chronological order to explain the sequence of my learning path. The later sections summarize practices and pitfalls that I learned across my career and will be useful to others undergoing digital transformations.

In this chapter, we'll cover the following main topics:

- A lifetime of study of testing, quality, security, and feedback for DevOps, DevSecOps, and SRE
- Bell-Northern – World-class university
- Testing as a commercial enterprise
- Consulting and teaching
- Lessons learned, pitfalls, and strategies to overcome pitfalls

Let's get started!

A lifetime of studying testing, quality, security, and feedback for DevOps, DevSecOps, and SRE

Going back, way back to the 1960s, I was a young boy growing up in a working-class family in Kingston, Ontario, Canada. We didn't have much money, but we had each other. My father and mother both worked hard to put food on the table and a roof over our heads for my two sisters and me. I was the baby of the family, with seven years between myself and my two older sisters. Essentially, I had three mothers because my sisters from the start saw me as "baby Marc," a moniker that stuck with me through adulthood.

This situation afforded me lots of time to play, and break things. This fact afforded me opportunities for experimentation because I was pampered and spoiled, which is good because I was born with a lazy disposition – an attribute that persists to this day and is a reason for my lifelong interest in automation.

I loved to take mechanical things and appliances apart and try to understand how they worked. These early experiences taught me a lot about the importance of testing, quality, security, and feedback because some of the things I took apart resulted in minor injuries and electric shocks for myself and my cat, Puff. (The cat was sometimes my lab assistant.) I also learned about the importance of following a quality process because I seldom could put things back together again without help from Dad.

Perhaps because they were not happy with things that I broke around the house, my parents, bought for me a series of MECCANO sets, which were kind of do-it-yourself kits made from mechanical metal parts – nuts, bolts, pre-punched struts and metal sheets, gears, wheels, belts, and pullies that you can create things with. *Figure 3.1* shows what this looks like.

Figure 3.1 – MECCANO

I was especially interested in automating the toys that I built using the MECCANO wind-up motor, and later the MECCANO electric motor. I remember building a loader that could drive itself. It would bump into the furniture because it had motion but no sensors or feedback. I was not satisfied with this because it had very limited utility compared to a real loader. Sensors were not available for MECCANO, so I was determined to look for another set of things to experiment with.

A few blocks from my house was a "magical" store called *The Hobby Shop*, which had low-cost toy models and many things that interested me. On my small weekly allowance, I was able to buy miniature plastic scale model airplanes, which not only could fly but also had control elements to steer the planes after they were launched with an elastic band. The control capability was very crude because it could not be adjusted after launching – a fact that also resulted in unintended mishaps, but I could adjust the flight path between test flights based on the feedback from prior flights. Model car racing on a track was also available at the model shop. This was more satisfying because I could adjust the speed with hand control, but the feedback was very limited from a visual point of view, so, too often the car would be overpowered and skip the track and crash. These early experiences taught me that timely feedback is needed to properly adjust controls and the combination of feedback and controls affect the quality and security of outcomes for even simple systems.

While building and operating models was fun, it did not satisfy my interest to have better control over the systems, and, they did not have day-to-day utility. My first bicycle was much more satisfying because I could at once control and monitor my direction and speed and use it to get somewhere, although it required a lot of effort to pedal. During my teen years, I graduated to building and operating motorized bikes, motorcycles, and cars, some of which are shown in *Figure 3.2*. Since money was tight, I would buy a bike or car that was not working for little money, and then work on it with my dad and mechanic friends to get it road-worthy. In the process, I learned a lot about mechanical and electrical systems and the importance of dashboard meters. Each transition exposed me to increasingly complex systems that required multiple control and feedback mechanisms to operate reliably and safely.

Figure 3.2 – Motorbikes and cars

I noticed that electronic components were used in many control and feedback functions. These were black boxes to me. So, I decided to study electronics in high school to understand what electronics is and how it works.

My first project in my grade 10 electronics class was to build a multimeter to measure current, voltage, and resistance. While it was a crude meter, it was accurate enough to use as the feedback meter for some early electrical and electronics projects, until I had enough money to buy a professional Fluke multimeter. With some money from summer jobs, and profit from selling my motorcycle and cars, I was able to purchase *Heathkits*, which were do-it-yourself kits for sophisticated electronic systems. Following the kit instructions, I built electronic tools, including a variable power supply, frequency counter, oscilloscope, and breadboard, as shown in *Figure 3.3*. These were sufficiently reliable and accurate to use throughout my engineering school years and beyond. (I still have the multimeter and power supply – it is still working, 50 years later).

Figure 3.3 – Electronics tools and Heathkits

I realized that, while electronics can perform many functions for sensing and control of complex systems, either people or automation is required to operate them to accomplish useful tasks. This is when my interest in computers and programming developed.

My first programming experience was in grade 10 in high school. We had access to a remote time-share mainframe computer to execute math formulas programmed in the **A Programming Language** (**APL**). The inputs (controls) and outputs (feedback) were in the form of an electronic terminal that looked like a typewriter, as shown in *Figure 3.4*. I liked the quick interactive response of the terminal, but I learned APL was limited to calculations and not as capable for general-purpose computing applications. We also had local access to a university mainframe computer with a card reader that could accept Fortran programs in the form of a stack of punched cards. The time-shared mainframe IBM360 would run each program as a batch job sequentially, and output (feedback) was generated on a line printer. Given the serial and variable nature of the read, run, and output, sequences' feedback could take many hours depending on the size of the program and the number of batch jobs queued. The time between program creation and feedback was frustrating and slow.

Figure 3.4 – Early computers and programming

A few years later, I purchased a Radio Shack TRS-80 personal computer and a Commodore 64 that was programmed directly from the keyboard in the Basic language. The micro-processer was dedicated to one single application. A cassette tape provided "bulk data" storage and retrieval because memory circuits were very expensive and limited capacity. A character-mode display would output characters in real time, which was nice for getting real-time feedback, but output formats were limited especially for graphs and charts.

These personal computers also had an acoustic coupler that would enable connection to a telephone handset for data transmission to other remote computers so I could experiment with data communications between the TRS80 and Commodore. This was my first experience with automated remote control and real-time remote feedback. This opened up possibilities for many new applications, for interactive remote control and feedback such as remote control of machines and interactive video games. But the communication speed was limited to 300 baud, by the modem. I soon realized that precise real-time control and feedback would only be accomplished with special hardware and firmware that could be built to more accurate machine-level timings that are needed in real-time control applications. I purchased some microprocessor evaluation kits and learned assembler and integrated circuits logic design to learn about real-time transmission control and feedback methods.

During my *Queen's University engineering* years, I learned a lot more about the importance of testing, quality, security, and feedback for system design.

For one of my summer jobs, I worked in the input/output department of an **Esso Oil** corporation's data center in Toronto that had a large IBM360 mainframe for computing logistics for Canada-wide oil deliveries. The input was a stack of punched cards, and the output was a "high speed" line printer that generated continuous reams of multi-copy paper separated by carbon papers. After the print was generated, the carbon paper and printed copies had to be separated by a "decollator" machine, and then the individual pages had to be separated and stacked using a "burster" machine. Then, the separate copies were bound in book fashion with metal pins to be suitable for reading by the analysts who received their copies in designated mail slots each morning.

The feedback time from card input to output (printed book) was routinely more than 16 hours for all the outputs needed for each day's logistics. If the analysts found a problem, it would have to go back in the queue for the next day's run. Problems with logistics meant that some deliveries of oil would be late because they needed the schedule to operate each day. Multiple times, especially when there was an errant system upgrade, an entire day's run tested bad, and the operations of logistics were interrupted for the entire data center. This again taught me that accelerating testing, quality assurance, security, and feedback was important to ensuring operations.

During my third-year summer, I was fortunate to get a job at the telecommunications manufacturer, **Northern Telecom (NT)**, who put me on assignment to **Bell-Northern Research (BNR)** because they needed a way to test the new state-of-the-art packet switching system that BNR was researching and developing. At BNR, under the direction of expert scientists, I developed special circuits, nano-code, micro-code, and high-level software for testing the trunk cards of the new packet switch. This cut the testing time from days to minutes for the technicians in manufacturing, greatly improved the precision of quality inspection, and reduced debugging time. This was a lesson that showed me the game-changing power of automation to improve feedback.

My Queen's University engineering thesis project, called **Intellicom**, was a collaboration with two of my engineering classmates. We designed and built one of the first space-switched multi-functional private office phone systems, shown in *Figure 3.5*. This ultimately resulted in recognition at a regional university engineering event.

Figure 3.5 – Intellicom engineering thesis project

We built the central controller, which consisted of a switch controller using a microprocessor board, a memory card, and telephone line cards with hybrid AD converters, all interconnected via a wire-wrapped backplane. The central controller and line cards had LED displays to provide real-time operational feedback. We also designed and built four multi-functional handsets with push-buttons for numbers and controls, and LEDs for end-user feedback. The control and feedback information between the central office unit and handsets was transmitted using control wires separate from the voice wires. In retrospect, this was a particularly early implementation of common channel signaling that later came to be the direction for telephone systems! Given a very limited student budget, we scrounged parts, such as handsets borrowed from our dormitory, and my summer student boss from BNR generously donated some parts from BNR's scrap piles. We had to make do with 256-byte eROM to hold the bootloader, and 1 Kbyte of 2,102 RAM for the entire operating software written in 68 K assembler code. This required some creative programming.

This was an ambitious project considering it had to be completed in time for graduation despite a very full course load! We realized from the start that the only hope of getting this working in time was to make sure our progress was always in the forward direction. Not much time to recover if there were big mistakes. So, we adopted an incremental design, build, test, lockdown approach – like what would later be called **Agile** and **DevOps**. As more and more parts were locked down, the test phase increased in complexity and time duration, a lot, so we automated the tests. This gave us back major time savings and improved confidence and quality immensely. Through destructive testing, we discovered and resolved potential security problems such as cross-talk and unauthorized listening. The most satisfying moment came when our professor saw all functions working and was unable to make our system crash during the final "demo day.".

BNR – World-class university

After graduation in 1978, I was hired by NT, in Belleville, Ontario, Canada, and initially assigned by NT to work at **Bell-Northern Research (BNR)**, in Ottawa, Ontario, Canada. My mission was to continue the work from my prior summer job, namely, to transfer technology of the experimental SL-10 packet switching system technologies from BNR labs to NT manufacturing.

In the late 1970s, BNR was the perfect learning environment. It was staffed with the top minds in the communications industry. The entire organization was extremely creative, energetic, innovative, and collaborative. Its culture was centered on telecommunications systems research and development. BNR had a truly amazing lab capability and funding was readily available for many projects. Working there presented many opportunities to collaborate with excellent scientists and engineers in the tri-corporate NT (later known as NorTel), BNR, and operating company **Bell Canada**.

I was assigned to the systems test team within the BNR packet switching group, while my NT role was to ensure the SL-10 was prepared for NT manufacturing to build and test SL-10 nodes and components in the factory. Included in this role was the development of a suite of switch-resident test programs to verify various system components called **In-Service Off-Line Testing (ISLOT)**. This unique set of responsibilities allowed me to learn about research, technology development, and business manufacturing operations. Since I was one of the few people in the organization with such a system-level role, I was also positioned to help with customer-facing demonstrations and interactions.

The early success of the SL-10 in field experiments gained recognition and NT received an order for an SL-10 from the *Deutsche Bundespost* in Germany, a project called **Berlin Packet Experiment** (**BERPEX**).

I was assigned to go to Berlin to install the switch, as shown in *Figure 3.6*. Since NT and BNR had no support offices in Germany, I had to be prepared to handle all problems locally, with the guidance of remote calls back to the scientists and engineers in BNR, Ottawa, Canada. The quality of my work and feedback to BNR was essential to the project's success, and the business expectations were very high. The customer would purchase a lot of SL-10 nodes for Germany's packet network if the Berlin node performed well. This was strategically important to NT, who wanted to expand their business into Europe.

I expected some issues, but it turned out to be a real "trial by fire." There were many problems found with the SL-10 node on-site in Berlin. We found and resolved power problems, connector, and cabling mismatches, software performance issues, power recovery procedures, problems with system commands to control system configurations, status and fault indicators, and protocol communication problems.

In addition, we discovered problems with communicating failure information, uploading remote fixes for testing, and difficulties with testing to verify fixes in the field. It took nine weeks, but finally, the switch was running well enough to satisfy the field trial goals of the experiment. Ultimately, Germany did proceed with additional orders, which became the first large data packet network in Europe called **DatexP**. In the process, we learned a lot about what it takes to get SL-10 nodes commissioned, what feedback mechanisms were needed for remote support, and what needed to be improved to speed up the process of commissioning and operations.

From a testing point of view, data protocol simulation using a microprocessor protocol tester designed by BNR and a data-monitoring tool called **DataSope** designed by NT were most important. These tools substituted for hosts and terminals during the installation and helped to diagnose connection problems at the various protocol layers. The results of tests could be printed and then relayed to the developer to diagnose problems and verify the subsequent fixes. Other feedback mechanisms were LED indicator lights on the circuit boards and the SL-10 character-mode video display administration terminal. Each of these tools provided valuable clues. They were not inter-connected and required a lot of manual coordination to formulate a coherent record of events. I realized that the testing, diagnostics, repair, and retesting cycle could be greatly improved if the tools were designed to be more integrated with the system and each other.

Figure 3.6 – SL-10 BERPEX

The successful experience of BERPEX signaled to BNR and NT that SL-10 was a major business opportunity. Already, we had new business prospects for packet networking coming from clients such as the Belgium bank, **Societe Generale de Banque (SGB)**, who wanted a network to be installed in 1979. But the quality needed improvement before we could deliver.

Returning from Berlin in early 1979, I was assigned to lead a testing team to qualify "Generic 10" of SL-10, which consolidated hardware and software changes determined during the Berlin experiment. It was optimistically estimated to take eight weeks to validate the system changes and reach our quality goal, including protocols and network configurations that the client required. However, the testing was manual and there were many thousands of tests. Each protocol required hundreds of tests to validate each update. We kept finding serious problems that reset the system test process. We discovered many problems with protocols that were not tested in Berlin and problems with multi-node test topologies that were beyond the scope of the Berlin project. The actual time to reach our quality goal for SL-10 Generic 10 was eight months – 6 months longer than the original plan! In the end, we had a switch that met the customer quality goals and we could ship to SGB.

I was assigned to commission the initial three SL-10 nodes on-site – the main node in Brussels, and two others in Antwerp and Ghent. The investment in SL-10 Generic 10 testing and quality improvements paid off. The node installations went much faster than with Berlin. I was able to commission each node and protocol interface in only a few days each, much reduced from the nine weeks experienced in Germany. The connections between nodes did take much longer but that was not the fault of NT or SL-10. The trunk links from the Belgium telephone company were delivered late. Once the trunk links were available, the SL-10 packet network came up and worked within a few days. SGB became a major client for NT and SL-10 became the packet system of choice for many other clients globally. I attribute part of the success on-site to the automated ISOLT test software and the trunk testing system that I developed before.

However, in retrospect, one thing was not good. The 6 month "over-estimate" was a problem. While there was general praise for the quality result, management was not happy with the time it took to qualify the release. I was given a mandate to figure out how to accelerate the time to quality for SL-10 from 8 months to 2 months.

The answer was obvious – automate the time-consuming tests and establish a configurable test lab that could accommodate the many different system topologies that different customers required. I was given the role of creating a new team for test tools development. Given prior lessons learned, I knew that the test system had to be powerful, scalable, and feature rich. It had to be something easy for developers, testers, and field staff to use so they could communicate tests and test results with each other efficiently. It had to be readily available with capacity and performance sufficient to test a high-performance, highly scalable packet network with many protocols, line, and trunk interfaces.

Over a few years, my team developed a comprehensive set of tools called **Network Test Environment (NTE)**. My paper, *An Integrated Test Center for SL-10 Packet Networks*, published in ACM journal, 1985, explained how it all worked. *Figure 3.7* shows some of the illustrations of the architecture of the primary tools in the set.

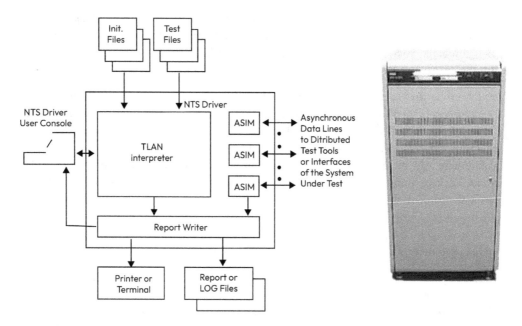

Figure 3.7 – Network test environment

There were four types of tools:

- **Interactive Protocol Tester (IPT)**: This simulated communication protocols and could be operated interactively or remotely through a CLI.

- **Network Load Test System (NLTS)**: This generated and measured network traffic.

- **Network Process Monitor (NPM)**: This could generate internal traffic, and perform network monitoring.

- **Network Test System (NTS)**: This was a general-purpose automated test controller that could automate actions and monitor results across a network of other tools and admin interfaces using a test language we developed called **Test Language (TLAN)**.

All these tools, except NTS, ran on the product hardware, which allowed them to scale to large configurations as well as the product itself and to offer use cases where the tools were integrated with the nodes on-site. This approach also ensured that the tools shared common administration and update processes. NTS operated on a commercial Unix-based microcomputer that was available from micro to mainframe size, so it could scale according to the test requirements and be serviced by the vendor. Since the NTS computer was the same make as the SL-10 data control center, it could be integrated and maintained with the same procedures as the product.

With these tools in place, we had a fully integrated set of testing tools capable of automating all tests in the lab, field trials, and on-premises. This was a major advantage compared to competitors. This contributed to SL-10 and the next-generation **Data Packet Network (DPN)**, achieving the highest level of efficiencies, quality, and sales worldwide in the industry of connection-oriented packet switching throughout the 1980s and early 90s.

The NTE tools gained industry recognition, continued to be upgraded, and became an important part of the global telecommunications network transformation from circuit-switching to **Signaling System 7 (SS7)** and **Integrated Services Digital Network (ISDN)**.

Along the way, I made leading contributions towards a global **International Organization for Standardization (ISO)** standards initiative to standardize protocol conformance testing with a notation and associated platform called **Tree and Tabular Combined Notation (TTCN)**.

As a result of this success, my career focus on network testing technologies was consolidated and enhanced with a series of projects in BNR, and then ECI Telecom and vPacket Communications, shown in *Figure 3.8*.

Figure 3.8 – Integrated network test technology

At BNR, I was promoted to director of global test technology responsible for creating test systems for all BNR and NT network and terminal products. From the mid-80s until 1991, my team created a series of testing systems for all the switching division products and third parties who developed terminal equipment to go with our products. These included **Test and Traffic Simulation (TATS)** automation platforms for network product system and regression testing, remote conformance testing platforms with standardized test suites for 3rd party qualification, and tools to measure test coverage.

One of the more important tool platforms that we developed called **Test Execution Administration Manager (TEAM)** reduced the time to set up and run network test cases by so much that it saved more than 90 million dollars per quarter – for years to come! It was a platform essential to our development, integration, and system testing processes across the company. TEAM was operated from a window in the developer's Unix workstations and provided control and feedback. It ensured each user session was secure and coordinated test environment reservations between the thousands of developers and lab systems required for network development and system testing.

Later at ECI Telecom and then vPacket, I directed the development of flexible remote access products that offered many configuration options that needed to be tested. To reduce the amount of time for testing, we built flexible test setups that could be reconfigured quickly while running automated test cases and designed-in test capabilities to monitor system performance built into the network products. This idea of integrating test capabilities seamlessly into the products and processes provided fast qualification of changes and improved collaboration during development, integration, and production.

Testing as a commercial enterprise

Since I had spent so much time in my career developing test technologies, it made sense for me to get involved in test systems as a commercial business. As shown in *Figure 3.9*, during the 1990s and early 2000s, I held various senior engineering and business roles over commercial test product platform development at **Tekelec**, **VIEW Engineering**, and **Spirent Communications**.

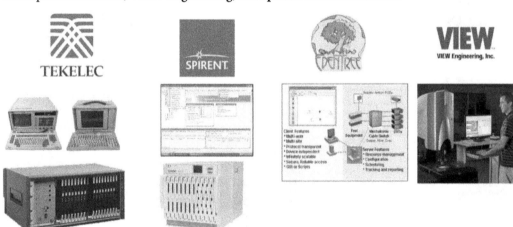

Figure 3.9 – Commercial test products

During this time, I founded a firm called *EdenTree Technologies* to produce smart test automation platforms for network laboratories. This exposed me to the testing and quality assurance environments of hundreds of global clients in networks, enterprises, operating companies, and government institutions. I found that testing was a bottleneck to the release processes and presented gating factors to achieve quality for most industries globally. I could see clearly that many of the lessons I had learned about testing, quality, feedback, and security previously in my career would help many organizations globally.

Consulting and teaching

From 2008, my career increasingly focused on providing senior consulting expertise for customers' testing, quality, feedback, and security processes, in support of their digital transformation projects for DevOps, DevSecOps, and SRE.

As a senior solutions architect at Spirent Communications, I advised clients on integrating test tools and automation with DevOps test environments and processes. I also advised on the development of a powerful test environment automation platform called **Velocity**.

At **Trace3**, I helped build the *Total DevOps* consulting practice that emphasized automating **continuous integration and continuous delivery (CI/CD)** processes for enterprise clients.

In 2019, I founded a boutique consulting firm called *Engineering DevOps Consulting* that offers prescriptive practices for DevOps, DevSecOps, and SRE transformations based on my book, *Engineering DevOps*, self-published with *Book Baby* in 2019.

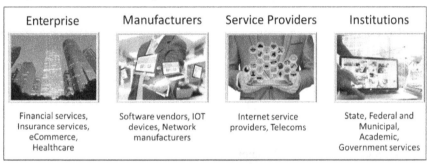

Figure 3.10 – Partners and Clients

As shown in *Figure 3.10*, today I provide DevOps, DevSecOps, and SRE consulting services for many enterprises, manufacturers, service providers, and institutions through partnerships with other consulting, training, and media firms such as the DevOps Institute, DevOps.com, Xellentro, Learning Tree International, and Opus Technologies.

Lessons learned, pitfalls, and strategies to overcome pitfalls

Looking back over my career experiences, which were described earlier in this chapter, I identified common pitfalls and strategies to overcome those pitfalls. The following sections highlight some of my learnings from the point of view of quality, testing, test automation, standardization, security, and feedback.

The importance of quality

The primary lesson I learned about quality is that too many organizations have a too narrow understanding of how to define, measure, and achieve the level of quality appropriate for their mission. Dr. Edward Deming, a renowned quality management expert, defined quality as a continuous process rather than a static attribute. He emphasized that quality is not merely about meeting specifications or having a product free of defects but involves meeting customer needs and exceeding their expectations.

The pursuit of "zero defects" is not an appropriate goal and pursuing this is wasteful. Instead, it is necessary to define use cases that are important to all the customers and stakeholders of the organization's mission. Good enough quality occurs when those use cases are fully satisfied.

Having a process for continuous quality evaluation and improvement is more important than achieving any one quality milestone because stakeholder requirements will evolve as a product or service matures.

Continuous testing is a key part of achieving a successful continuous quality program because testing provides evidence of compliance with stakeholder use cases. However, testing alone is not enough. Other factors, such as responsiveness to support requests, ease of use, integration, cost effectiveness, and innovative product roadmaps are equally important to meeting customer quality expectations and need to be monitored for an organization to continue to be successful.

Building testing tools into systems

I learned that testing is an essential and major part of every process, product, and service. Testing activities include test planning, test creation, test environment setup, coordination, test execution, results reporting, results analysis, fault reporting, and repair retest. Testing is often responsible for more than 50% of the total cost and time of the value stream. Successful products and services recognize this by building testing and monitoring capabilities in their products and processes. In this way, the testing capabilities are readily available when needed and can scale as the product scales. This also ensures test data and results are fed back to all the stakeholders in a form that they can immediately use to collaborate, which reduces waste.

Test automation for efficiency and competitiveness

In most cases, it is better to fully automate all types of tests. Tests need to be repeated often throughout each pass, and each stage, of a product's value stream. Without automation, every stage will be delayed waiting for someone to run the tests. In addition, manual testing is subject to human error, resulting in wasteful, false positive and false negative results.

The most competitive products prioritize test automation because the time efficiencies and quality benefits derived from automated testing give an organization a strategic advantage over competitors that don't.

However, you must be careful to have strong standards for test automation, or else test maintenance and results analysis become a bottleneck and excessive rework.

Advances with AI/ML help improve the quality and speed of every aspect of test automation and maintenance.

Standards accelerate collaboration

Without strong standards for all test artifacts, test automation can get out of control and become wasteful. Standards facilitate collaboration between tools and team members. It is important to have standards for quality, test strategies, test plans, test cases, test scripts, test environments, test execution, test results, test analysis, and feedback reports.

It is also important to define **service-level objectives** (**SLOs**) that govern how test failures are prioritized and handled by the organization, or else the process can cause organization friction.

Implementation of a common test engineering platform made available as a service to development, test, and operations staff is ideal to enable communication of testing artifacts and to ensure everyone in the organization has a common "source of truth" for testing.

Security requires a comprehensive approach

Cybersecurity concerns are continuously evolving and escalating as systems become more distributed, delivered using multiple pipelines, and deployed over distributed ephemeral infrastructures. Security becomes a major concern and bottleneck in development, integration, delivery, and production unless security controls are built into the end-to-end value stream.

A continuous security mindset and implementation of security practices requires a combination of DevSecOps practices and in-production SecOps practices to ensure security concerns are minimized pre- and post-deployment to production. The entire development team needs secure coding training and security practices need to be led by threat modeling appropriate for each application. Code, third-party libraries, and image scanners during integration and delivery need to be tuned according to the most likely threats.

Pen testing and chaos security engineering need to be automated as part of delivery processes before release to production. All processes and assets need to be protected with identify-based zero-trust technologies and firewalls because security threats can come from internal or external sources to the organization. Deployments are best using immutable technologies such **infrastructure as code (IaC)**, containers, and container orchestration tooling that enable infrastructure to respond rapidly to changing threat conditions. Monitoring of security events and indicators needs to be visible throughout each value stream from ideation to production.

Without feedback, you are running blind

Feedback in the form of a combination of quality metrics, testing metrics, and security metrics is critical to monitor and manage the efficient flow of changes for development, integration, delivery, and deployment. Thresholds and goals can be defined as SLOs, and measured continuously using **service-level indicators (SLIs)**. Error budgets and error budget policies, concepts from SRE practice, are useful to manage these activities, to ensure the metrics are respected by the organization.

Summary

This chapter reflected on my career experiences to distill insights into common pitfalls and effective strategies in quality, testing, security, and feedback. It began by emphasizing the multifaceted nature of quality, echoing Deming's perspective that quality is not a static attribute, but an ongoing process aligned with customer needs and exceeding their expectations. The pursuit of "zero defects" was deemed wasteful, highlighting the importance of identifying crucial use cases and achieving satisfactory quality through their fulfillment. Continuous evaluation and improvement were emphasized, recognizing the evolving nature of stakeholder requirements.

Testing emerged as a pivotal aspect, consuming a significant portion of resources in the value stream. Successful products integrate testing capabilities within the product itself, ensuring scalability and immediate availability of test data for collaboration among stakeholders, thereby reducing waste.

Test automation stood out as a driver of efficiency and competitiveness. The chapter advocated for comprehensive automation to avoid delays and errors inherent in manual testing but stressed the necessity of strong standards to maintain efficiency in test maintenance and results analysis.

The establishment of standards was highlighted as crucial for collaboration and control in testing practices. From quality to feedback reports, standards were deemed essential. A common test engineering platform was advocated for, which enhanced communication and a shared foundation of truth across the organization.

Addressing security concerns, the chapter underscored the need for a comprehensive approach throughout the value stream. From DevSecOps practices to in-production SecOps practices, the integration of security measures was emphasized. Threat modeling, automated security testing, and a zero-trust approach were recommended to mitigate both internal and external security threats.

Lastly, the importance of feedback mechanisms, incorporating quality, testing, and security metrics, was stressed. SLOs, SLIs, and error budget policies were highlighted as practical tools for continuous monitoring and management, ensuring organizational alignment with defined metrics.

The next chapter will explain an engineering approach to guide transformation activities toward success.

Part 2: Determining Solutions Priorities

Part 2 of this book focuses on the strategic planning and prioritization necessary for effectively implementing continuous practices in an organization. It begins with an exploration of a systematic engineering approach to planning solutions for continuous testing, quality, security, and feedback. This sets a disciplined framework for organizations to follow, ensuring that their transformation efforts are both structured and effective.

The section progresses by discussing how to determine transformation goals that are tailored to the specific needs of different organizations, products, and services. It introduces tools and methodologies that help in setting these goals, providing a clear roadmap for businesses to follow. Additionally, the section covers the importance of understanding the current state of an organization's capabilities through discovery and benchmarking, which is crucial for identifying areas that need improvement. The selection of appropriate tool platforms and the integration of AI and ML technologies are also discussed, highlighting how these can enhance continuous practices and foster a culture of continuous improvement and resilience. This part of the book is essential for anyone looking to strategically prioritize and streamline their approach to digital transformation through continuous practices.

This part includes the following chapters:

- *Chapter 4, Engineering Approach to Continuous Testing, Quality, Security, and Feedback*
- *Chapter 5, Determining Transformation Goals*
- *Chapter 6, Discovery and Benchmarking*
- *Chapter 7, Selecting Tool Platforms and Tools*
- *Chapter 8, Applying AL/ML to Continuous Testing, Quality, Security, and Feedback*

4

Engineering Approach to Continuous Testing, Quality, Security, and Feedback

This chapter explains a systematic, disciplined engineering approach to planning continuous testing, quality, security, and feedback solutions. This approach was developed over the years in response to experiences that were described in the previous chapter. It includes the Seven-Step Transformation Engineering Blueprint, explains the importance of an expert transformation consultant and the use of AI tools to accelerate the work, and an explanation of **Capability Maturity Models (CMMs)** and levels of capability maturity for continuous testing, quality, security, and feedback.

In this chapter, we'll cover the following main topics:

- Why is an engineering approach needed?
- Seven-Step Transformation Engineering Blueprint
- Capability maturity models guide transformation
- Capability maturity levels – Continuous testing
- Capability maturity levels – Continuous quality
- Capability maturity levels – Continuous security
- Capability maturity levels – Continuous feedback

Let's get started!

Why is an engineering approach needed?

Implementing continuous testing, quality, security, and feedback requires a well-engineered digital transformation approach because it's not just about adopting new tools but fundamentally changing how an organization operates, delivers, and measures value. This involves redefining processes, culture, and mindsets to integrate these elements seamlessly into the software development life cycle. It's a holistic change, requiring a shift in practices and collaboration across departments, and aligning these efforts with the broader business objectives. This transformation ensures that quality and security are not afterthoughts but are integral, continuously monitored, and continuously improved aspects of the development process.

A digital transformation for an organization impacts not only technology but also strategy, operations, and customer engagement. The goal is to improve efficiency, manage risk, and discover new monetization opportunities.

Organizations often face challenges in their digital transformations due to several key pitfalls:

- **Lack of clear vision and strategy**: Without a clear direction and understanding of what digital transformation entails, organizations can struggle to align their efforts effectively.

- **Resistance to change**: Cultural resistance within the organization, particularly from employees accustomed to traditional ways of working, can hinder the adoption of new technologies and processes.

- **Inadequate leadership commitment**: Transformation requires strong leadership to drive change, overcome resistance, and allocate necessary resources.

- **Poor planning and execution**: Inadequate planning, unrealistic timelines, and insufficient resources can lead to failed implementation.

- **Technology challenges**: Over-reliance on technology without considering the necessary process and culture changes can lead to suboptimal outcomes.

- **Failure to scale**: Difficulty in scaling successful pilots to the broader organization can impede transformation efforts.

- **Inadequate skills and expertise**: Lack of necessary skills among employees to adapt to new digital tools and processes can be a significant barrier.

- **Lack of measurements**: Not having metrics to monitor progress, performance, and **Return on Investment** (**ROI**) can lead to misaligned efforts and an inability to gauge success or failure accurately.

- **Inadequate service and sustainability**: Failing to ensure that digital solutions are maintainable and sustainable over time can result in systems that are not viable in the long term, undermining the entire transformation effort.

An engineering approach is crucial for successful digital transformations because it brings a structured, disciplined methodology to what is often a complex, multifaceted process. This approach ensures that digital transformation is not just about adopting new technologies but about re-engineering processes, workflows, and organizational structures. It helps in methodically addressing the technical, process, and cultural aspects of transformation, reducing risks, improving efficiency, and enhancing the likelihood of achieving desired outcomes. By applying engineering principles, organizations can systematically plan, implement, measure, and iterate their transformation efforts, aligning them with business goals and ensuring sustainable, long-term success.

Understanding the Seven-Step Transformation Engineering Blueprint

The **Seven-Step Transformation Engineering Blueprint** shown in *Figure 4.1* is described in my book *Engineering DevOps*. The blueprint prescribes an infinite cycle of seven steps for achieving your transformation goals methodically, no matter what your goals or level of maturity are currently. The same seven steps apply to organizations that want to transform their testing, quality, security, and feedback processes into continuous processes.

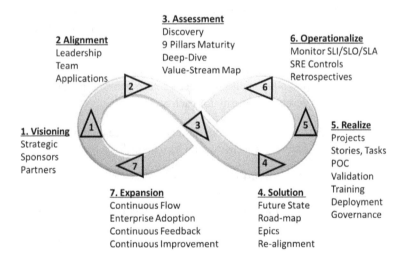

Figure 4.1 – Seven-Step Transformation Engineering Blueprint

Transforming from one maturity level to the next higher level requires transitioning through the seven steps of each improvement cycle to effect progressive refinements necessary for a solution that maintains a balance across three dimensions of transformation (people, processes, and technology) and pillars of practices for continuous testing, quality, security, and feedback, as described in *Chapter 2*. When the dimensions and pillars are out of balance, friction will occur and efficient flow in the pipeline and goals will be impacted. Changes to the dimensions and pillars must be introduced incrementally and tested methodically for each cycle in a continuous improvement loop.

The **Seven-Step Transformation Engineering Blueprint** includes the following seven steps:

1. **Visioning**: Define the strategic needs of the organization and identify sponsors that will own the transformation at the strategic level, as well as key partner organizations that need to be strategically aligned to the transformation.

2. **Alignment**: Leaders and key stakeholders that are most important to each transformation cycle agree to specific goals for selected applications.

3. **Assessment**: For the current state of selected applications, maturity is discovered and assessed, deep-dive assessments are conducted for specific topics, and a current-state value-stream map is created relative to the organization's goals.

4. **Solution**: An expert team performs an analysis of assessment data and formulates a future-state value-stream roadmap, including themes, epics, and user stories, and obtains alignments with key stakeholders.

5. **Realize**: Implementation projects are defined including user stories and tasks, product **Proof of Concept** (**PoC**) trials are conducted to validate solution choices, the solution is validated with selected applications and use cases, training is conducted as the solution is deployed to production, and governance practices for the new solution are activated.

6. **Operationalize**: Deployed improvements are monitored and controlled with **Site Reliability Engineering** (**SRE**) practices that monitor SLI, SLO, and SLA metrics. Retrospectives are conducted to create actionable prioritized lessons learned for continuous improvement.

7. **Expansion**: Once the solution is realized for a select set of applications, the organization can safely expand the solutions to other applications across the organization. Further transformation cycles will lead to higher levels of maturity for each application.

This engineering blueprint is crucial for organizations aiming to enhance their practices to higher levels of capability maturity, ensuring a balanced, systematic, and incremental approach to their transformation.

If an enterprise does not adhere to this transformation blueprint, several risks and consequences may arise, undermining the efficiency, effectiveness, and overall success of their transformation. Here are the risks and consequences associated with not following the blueprint steps:

1. **Visioning**:

    -   **Risk**: Without a clear vision and well-documented strategic need for the transformation, the organization may lack direction and purpose in its transformation efforts. This can lead to misaligned objectives and wasted resources.

    -   **Consequence**: The absence of strong sponsorship and strategic alignment with key partners may result in resistance to change, low adoption rates among teams, and a failure to realize the full benefits of the transformation.

2. **Alignment**:

   - **Risk**: Failing to achieve alignment on goals for transformations can lead to inconsistencies in expectations and deliverables across different teams and departments.

   - **Consequence**: This misalignment can cause friction between teams, inefficient workflows, and a lack of collaboration, all of which can delay or derail transformation initiatives.

3. **Assessment**:

   - **Risk**: Skipping the assessment of the current state and capability maturity can leave gaps in understanding the baseline from which improvements are to be made.

   - **Consequence**: Without a clear assessment, it's challenging to measure progress or identify areas needing improvement, leading to the inefficient use of resources and potentially missing critical issues that could impede transformation success.

4. **Solution**:

   - **Risk**: Not leveraging expert analysis to formulate a future-state roadmap may result in poor choices for the solution and a lack of clarity and direction for the transformation journey.

   - **Consequence**: The organization might implement solutions that are not aligned with its specific needs, leading to ineffective practices and increased costs and potentially failing to achieve desired outcomes.

5. **Realize**:

   - **Risk**: Overlooking the realization phase, including POC trials and validation, can lead to deploying solutions that are not fully tested or understood.

   - **Consequence**: This can result in operational disruptions, degraded user experience, decreased productivity, and a lack of confidence in practices among team members.

6. **Operationalize**:

   - **Risk**: Ignoring the need to monitor and control deployed improvements can lead to a lack of visibility into the performance and effectiveness of practices.

   - **Consequence**: Without proper monitoring and retrospectives, the organization misses opportunities for continuous improvement, potentially leading to stagnation and inefficiency in operations.

7. **Expansion**:

   - **Risk**: Failing to expand successful practices across the organization can limit the transformation's impact to only a select set of applications or teams.

- **Consequence**: This limited scope can prevent the organization from achieving broader operational efficiencies, reducing the overall ROI in transformation efforts.

In summary, not adhering to the Seven-Step Transformation Engineering Blueprint can significantly hinder an organization's ability to successfully implement and benefit from practices. These risks highlight the importance of a structured, strategic approach to transformation, emphasizing the need for clear visioning, alignment, assessment, solution development, realization, operationalization, and expansion to ensure successful integration into the enterprise.

The next section explains how the seven steps toward transformation can be accelerated.

Expert and AI-accelerated transformations

As illustrated in *Figure 4.2*, it is recommended that an experienced transformation expert consultant, or a small team of experts, supported by AI tools, guide the transformation process, serving in the role as an "expert transformation consultant" either individually or collectively.

Figure 4.2 illustrates the central role of transformation consultants as an orchestrator of the transformation process. The transformation consultant provides tools and templates to help accelerate each of the steps of the transformation process. At each step in the process, the tools and templates can be customized to tailor to the needs, maturity level, and preferences of the organization. This process can be accelerated using AI tools at each step to suggest customizations for templates relevant to each step based on results from the prior steps.

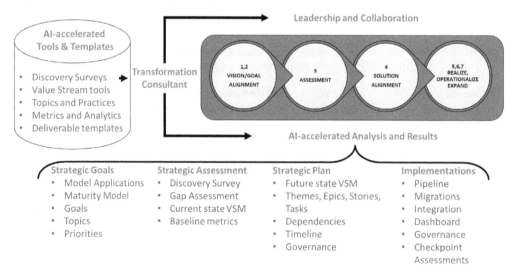

Figure 4.2 – AI-accelerated transformation process

The role of an expert consultant in orchestrating the Seven-Step Transformation Engineering Blueprint is critical for ensuring the success of an organization's transformation efforts. These consultants bring specialized knowledge, experience, and skills to guide and facilitate each step of the transformation process, ensuring that strategic objectives are met efficiently and effectively.

AI tools, when guided by expert consultants, can significantly enhance the effectiveness, speed, and quality of each step in the transformation process.

The following is an outline of the need and role of an expert consultant and the use of AI tools at each step of the blueprint:

1. **Visioning**:

   - **Need**: Establish a clear and strategic direction for the transformation.

   - **Consultant role**: Expert consultants help define the organization's vision for the transformation, ensuring it aligns with its overall strategic objectives. They work with leaders to identify transformation sponsors and align key partner organizations, leveraging their experience to foresee and mitigate potential challenges.

   - **AI's role**: AI can analyze data from various sources to identify trends, opportunities, and threats that may not be immediately apparent. This analysis can inform the strategic direction of the transformation.

   - **Consultant's guidance**: Consultants can interpret AI-generated insights to align them with the organization's strategic objectives, ensuring that the vision for transformation is both ambitious and achievable.

2. **Alignment**:

   - **Need**: Ensure all stakeholders are working toward common goals.

   - **Consultant role**: Consultants facilitate alignment sessions among leaders and key team members, using their expertise to bridge gaps between different visions and objectives. They help define specific, achievable goals for the transformation, ensuring that everyone is committed to the same outcomes.

   - **AI's role**: AI-powered tools can facilitate stakeholder alignment by offering platforms for collaboration, goal-setting, and tracking. **Natural Language Processing** (**NLP**) can analyze communication patterns to identify misalignments or areas of confusion among stakeholders.

   - **Consultant's guidance**: Consultants can use these insights to mediate discussions, clarify objectives, and ensure that all parties are aligned toward common goals.

3. **Assessment**:

   - **Need**: Understand the current state and identify areas for improvement.

   - **Consultant role**: Through assessments, consultants diagnose the current maturity of the organization's processes and technologies. They conduct deep-dive analyses into specific topics and create value-stream maps, providing a detailed understanding of the existing state and identifying critical areas for transformation.

   - **AI's role**: AI can automate the assessment of current systems and processes by analyzing code bases, infrastructure configurations, and workflow patterns to identify bottlenecks or inefficiencies.

   - **Consultant's guidance**: Consultants can then interpret these findings, prioritize areas for improvement, and tailor the transformation strategy to address the most critical gaps first.

4. **Solution**:

   - **Need**: Design a roadmap for the future state that aligns with organizational goals.

   - **Consultant role**: Expert consultants analyze assessment data to develop a comprehensive future-state roadmap, including themes, epics, and user stories. They ensure stakeholder alignment and refine the plan to meet the unique needs of the organization, drawing on their experience to propose effective solutions.

   - **AI's role**: Using AI, consultants can simulate different transformation scenarios to predict outcomes, identify potential risks, and optimize the transformation roadmap.

   - **Consultant's guidance**: This allows consultants to present a data-driven future-state roadmap, ensuring solutions are both innovative and aligned with the organization's goals.

5. **Realize**:

   - **Need**: Implement the transformation effectively and ensure its alignment with strategic goals.

   - **Consultant role**: Consultants oversee the definition of implementation projects, conduct POC trials, and validate solution choices. They ensure that the solution is effectively deployed, training is conducted, and governance practices are established, leveraging their expertise to anticipate and address challenges.

   - **AI's role**: AI tools can streamline the implementation of transformation projects by automating routine tasks, optimizing resource allocation, and predicting the impact of changes before they are made.

   - **Consultant's guidance**: Consultants can leverage AI to ensure that the deployment of solutions is efficient, effective, and closely monitored, addressing any issues proactively.

6. **Operationalize**:

   - **Need**: Ensure that improvements are sustainable and monitored for continuous optimization.

   - **Consultant role**: By implementing SRE practices, consultants help monitor key performance indicators (SLIs, SLOs, and SLAs) to ensure that the deployed improvements are delivering the expected value. They facilitate retrospectives to derive lessons learned and prioritize them for continuous improvement.

   - **AI's role**: AI can enhance SRE practices by suggesting observability and advanced monitoring capabilities, predictive analytics for system behavior, and automated incident response.

   - **Consultant's guidance**: Consultants can use AI insights to fine-tune performance indicators, ensuring that improvements are sustainable and that the organization is continuously optimizing its operations.

7. **Expansion**:

   - **Need**: Scale the transformation across the organization to maximize its impact.

   - **Consultant role**: Expert consultants guide the organization through the process of expanding the successful practices to additional applications or teams. They help plan and execute further transformation cycles, ensuring that each cycle leads to higher levels of maturity and efficiency.

   - **AI's role**: AI can help identify which practices are most effective and ready for scaling by analyzing performance data across different applications and teams.

   - **Consultant's guidance**: Consultants can guide the strategic expansion of successful practices, using AI to ensure decisions are data-driven and that the organization is prepared for broader transformation.

Expert consultants play a pivotal role at each step of the transformation blueprint, offering the necessary guidance, expertise, and support to ensure that the transformation is strategic, aligned, effectively implemented, and sustainable. Their involvement helps organizations navigate the complexities of transformation, avoid common pitfalls, and achieve their desired outcomes more efficiently and effectively.

The transformation consultant can use AI tools to accelerate the creation of well-tailored analyses and documents resulting from each step of the transformation process.

In each step, the combination of AI tools and expert consultants creates a powerful synergy. AI provides the data-driven insights and automation capabilities to streamline and enhance decision-making processes, while consultants bring the strategic oversight, industry knowledge, and human judgment necessary to interpret AI outputs and guide the transformation process effectively. This collaborative approach ensures that the transformation is not only aligned with the organization's goals but also leverages the latest technology advancements to achieve superior outcomes.

Capability maturity models guide transformations

A CMM is a framework that helps organizations evaluate the effectiveness and maturity of their business processes. It's important for organizational transformations to have a maturity model, because it provides a structured path to guide improvements. CMM helps identify the current level of process maturity and guides organizations to advance through higher levels of discipline and quality. This structured approach ensures that changes are sustainable, processes are efficient, and the organization is better equipped to meet its goals and adapt to new challenges.

Capability Maturity Model Integration (**CMMI**) evolved from the original CMM, which was developed in the late 1980s by the **Software Engineering Institute** (**SEI**) at Carnegie Mellon University. CMM was initially focused on software development processes. Over time, the need for a more integrated approach became evident, leading to the development of CMMI in the early 2000s. CMMI expanded the scope beyond software to include other areas, such as product life cycle management and service delivery, offering a more detailed and flexible framework. This broader applicability and integrated approach distinguish CMMI from the original CMM.

Figure 4.3 illustrates the five maturity levels of CMMI.

Figure 4.3 – CMMI maturity levels

These levels are as follows:

- **Level 1: Initial** – Processes are unpredictable and reactive, often resulting in projects being delayed and going over budget.

- **Level 2: Managed** – Processes are managed on the project level, with projects being planned, performed, measured, and controlled.

- **Level 3: Defined** – The organization operates proactively with organization-wide standards providing guidance across projects, programs, and portfolios.

- **Level 4: Quantitatively Managed** – The organization is data-driven, with quantitative performance improvement objectives that are predictable and align with the needs of internal and external stakeholders.

- **Level 5: Optimizing** – The organization is stable and flexible, focused on continuous improvement, and able to respond to change and opportunities. It provides a platform for agility and innovation.

A modified CMM is necessary for software delivery capabilities because CMMI was primarily designed for more predictable and stable environments, which do not fully address the dynamic and iterative nature of modern software development. Today's software delivery requires rapid adaptation, **Continuous Integration (CI)** and delivery, and the ability to respond to changing market demands quickly. A modified model incorporates these aspects, emphasizing agility, automation, DevOps continuous delivery practices, and continuous learning, for the people, processes, and technology dimensions, which are crucial for the current landscape of software delivery.

Figure 4.4 outlines a software delivery CMM with five levels:

- **Level 1: Manual & Unmeasured (Chaos)** – Characterized by chaos, poor communication, manual processes, disparate tools, and lack of metrics:

  - **People**: Poor communication between team members. Blameful culture.

  - **Process**: Orchestration of integration and delivery processes are manual.

  - **Tech**: Disparate tools used by different teams. No integrated toolchain.

  - **Metrics**: Few, if any, metrics are automated. Releases are unpredictable and error prone.

- **Level 2: Continuous Integration (CI)** – Features well-engineered integration practices, collaborative culture within teams, orchestrated CI processes, coherent toolchains, and automated CI metrics:

  - **People**: Well-engineered integration practices in place for dev and QA teams. CI training in place. Collaborative culture within teams.

  - **Process**: CI processes are orchestrated and automated. Code is merged and tested to a common trunk branch during CI.

  - **Tech**: Coherent toolchain in place for CI version management, builds, and tests. Release candidates are managed in an artifact repo.

  - **Metrics**: CI metrics are automated and used to govern CI activities. Developers check in code daily.

- **Level 3: Continuous Delivery (CD1)** – Involves collaborative delivery practices across teams, automated delivery processes and metrics, and a coherent toolchain that includes automated testing tools:

    - **People**: Delivery practices in place for dev, QA, and release teams. Collaborative culture between teams.

    - **Process**: Delivery processes are orchestrated and automated. Tests for regression, system performance, and user acceptance are automated.

    - **Tech**: Coherent toolchain in place for CD, including automated tools for testing.

    - **Metrics**: Delivery metrics are automated and used to govern delivery and release acceptance activities.

- **Level 4: Continuous Deployment (CD2)** – Involves deployment practices across all teams, fully automated deployment processes, and metrics to govern and drive continuous improvement:

    - **People**: Deployment practices in place for dev, QA, release, and ops teams. Collaborative culture between teams.

    - **Process**: Deployment to production processes orchestrated and automated.

    - **Tech**: Coherent toolchain in place for deployments, including automated tools for infrastructure orchestration and in-prod monitoring and auto-roll-backs when needed.

    - **Metrics**: Deployment metrics are automated and used to govern deployment to production and drive continuous improvement.

- **Level 5: Continuous Operations (CO)** – Teams work collaboratively within a blameless culture, production operations processes are unified, and a unified toolchain is in place for full life cycle management, governed by SLAs, SLOs, SLIs, and error budgets:

    - **People**: Dev, QA, release, and ops teams work collaboratively within a blameless culture to ensure product quality and reliability meet stakeholder requirements.

    - **Process**: Production operations processes are unified between operations and development.

    - **Tech**: Unified toolchain in place for testing and monitoring applications, infrastructure, and release deployments.

    - **Metrics**: SLOs, SLIs, error budgets, and error budget policies govern development, delivery, deployment, and operations.

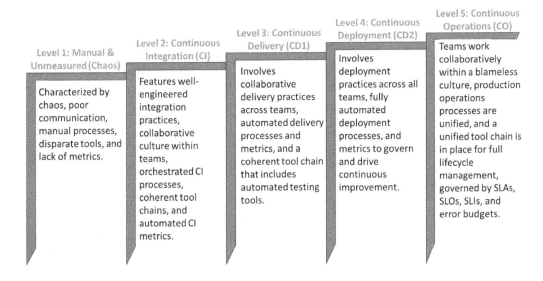

Figure 4.4 – Software delivery CMM

There is no standard software delivery CMM, but it is recommended that each organization determine a model suitable for gauging the maturity of that organization.

The following sections in this chapter provide examples of maturity models specific to continuous testing, quality, security, and feedback. While there is no standard model, the models presented in this chapter are consistent with the definitions provided in *Chapter 1* and the structure of the model illustrated in *Figure 4.4*.

Capability maturity levels – Continuous testing

The continuous testing CMM outlined in *Figure 4.5* describes five levels of maturity for continuous testing practices:

- **Level 1: Manual & Unmeasured (Chaos)** – Ad hoc and manual testing processes. Limited use of basic testing tools. Few if any structured metrics or KPIs for testing:

  - **People**: Poor communication between team members. Blameful culture.

  - **Process**: Orchestration of integration and delivery processes are manual.

  - **Tech**: Disparate tools used by different teams. No integrated toolchain.

  - **Metrics**: Few, if any, metrics are automated. Releases are unpredictable and error prone.

- **Level 2: Continuous Integration (CI)** – Initial automation of some test cases. Integration of testing in the CI pipeline. Basic metrics for test coverage and effectiveness:

  - **People**: Initial collaboration between developers and QA testers begins.

  - **Process**: Testing processes are defined and automated for CI.

  - **Technology**: Introduction of automated testing tools for CI test cases.

  - **Metrics**: Basic test coverage and defect metrics are tracked, at least for CI processes.

- **Level 3: Continuous Delivery (CD1)** – Full integration of testing into the CI/CD pipeline. Enhanced test automation capabilities. Regular use of metrics for test process optimization:

  - **People**: Clear roles for continuous testing practices are established.

  - **Process**: Testing is fully integrated into the development life cycle as needed to test a release candidate.

  - **Technology**: Use of a coherent set of tools for test automation.

  - **Metrics**: Regular use of metrics to improve test processes and coverage, for release candidates.

- **Level 4: Continuous Deployment (CD2)** – Advanced, automated, and streamlined testing processes. High-level test automation and orchestration tools. Detailed analytics for continuous testing improvement:

  - **People**: High level of collaboration across all team members in testing.

  - **Process**: Advanced, automated testing processes are standardized, including automated tests for deployment to production.

  - **Technology**: Sophisticated test automation and orchestration tools in place.

  - **Metrics**: Detailed metrics and KPIs for continuous improvement are used for deployment validation.

- **Level 5: Continuous Operations (CO)** – Organization-wide focus on quality and testing. Adaptive testing strategies using AI and ML. Predictive analytics for testing efficiency and effectiveness:

  - **People**: Organization-wide commitment to quality and continuous testing.

  - **Process**: Self-adaptive testing processes enabled by AI and machine learning.

  - **Technology**: Fully automated, self-learning testing systems.

  - **Metrics**: Predictive analytics used for continuous refinement of testing.

This model helps organizations assess and improve their continuous testing practices, aligning them with the broader goals of software delivery and operational excellence. Each level provides specific focus areas for advancement in testing processes, technology, and metrics, guiding organizations in their journey toward continuous testing maturity.

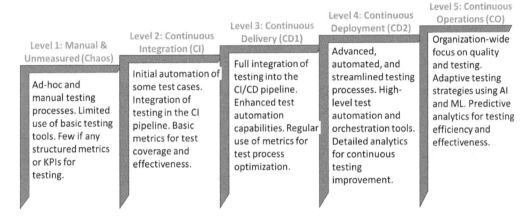

Figure 4.5 – Continuous testing CMM

This model is useful for organizations to evaluate and improve their continuous testing practices. It defines clear criteria for each level, allowing organizations to identify where they stand in terms of process maturity and technology integration. By assessing against these criteria, organizations can pinpoint specific areas for improvement and strategize their progression toward more advanced levels of testing maturity, which are crucial for achieving higher efficiency, quality, and speed in software delivery.

Capability maturity levels – Continuous quality

The model illustrated in *Figure 4.6* presents a continuous quality CMM with five levels:

- **Level 1: Manual & Unmeasured (Chaos)** – This level is characterized by limited communication about quality, manual quality checks, disparate non-integrated tools, and minimal quality metrics tracking:

  - **People**: Limited communication about quality. Reactive approach.

  - **Process**: Manual quality checks. No integration into the development life cycle.

  - **Technology**: Disparate, non-integrated quality tools.

  - **Metrics**: Few, if any, quality metrics. Minimal tracking.

- **Level 2: Continuous Integration (CI)** – Features some collaboration between development and QA teams on quality, automated quality checks during CI, basic automated testing tools integrated with CI, and basic quality metrics tracked during CI:

  - **People**: Some collaboration between development and QA teams on quality.

  - **Process**: Automated quality checks integrated into CI.

  - **Technology**: Basic automated testing tools integrated with CI.

  - **Metrics**: Basic quality metrics tracked during CI.

- **Level 3: Continuous Delivery (CD1)** – Enhanced team collaboration for quality assurance, fully integrated quality checks into CD processes, advanced automated testing and quality tools for CD, and comprehensive quality metrics tracked to guide CD decisions:

  - **People**: Enhanced collaboration across all teams for quality assurance.

  - **Process**: Quality checks fully integrated into CD processes.

  - **Technology**: Advanced automated testing and quality tools for CD.

  - **Metrics**: Comprehensive quality metrics tracked, guiding CD decisions.

- **Level 4: Continuous Deployment (CD2)** – A strong focus on quality in deployment practices, automated and streamlined quality processes for deployment, sophisticated tools to ensure quality in deployments, and detailed quality metrics for deployment and feedback loops:

  - **People**: Strong focus on quality in deployment practices.

  - **Process**: Automated and streamlined quality processes for deployment.

  - **Technology**: Sophisticated tools for ensuring quality in deployments.

  - **Metrics**: Detailed quality metrics for deployment and feedback loops.

- **Level 5: Continuous Operations (CO)** – An organization-wide commitment to quality in operations, unified quality processes between development and operations, integrated tools for monitoring and maintaining quality, and advanced metrics, including SLA, SLOs, SLIs, and error budgets, to govern quality:

  - **People**: Organization-wide commitment to quality in operations.

  - **Process**: Unified quality processes between development and operations.

  - **Technology**: Integrated tools for monitoring and maintaining quality.

  - **Metrics**: Advanced metrics, including SLOs and SLIs, to govern quality.

This model is useful for describing the maturity levels of an organization's continuous quality practices, aligning them with the broader goals of software delivery and operational excellence. Each level defines specific focus areas for advancing quality processes, technology, and metrics, guiding organizations in their journey toward continuous quality improvement. By evaluating their current state against the model's criteria, organizations can identify specific practices that need to be developed or enhanced to reach the next maturity level.

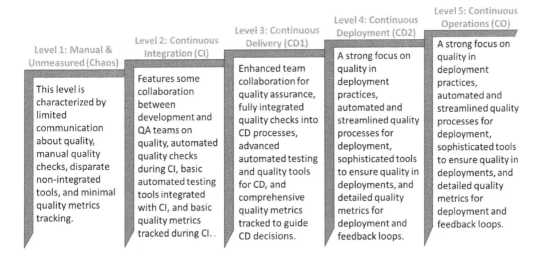

Figure 4.6 – Continuous quality CMM

While there is no standard continuous quality CMM, each organization should define one suitable for the organization. The model provided in *Figure 4.6* can be used as a basis for creating one specific to the organization.

Capability maturity levels – Continuous security

Figure 4.7 presents a continuous security CMM with five levels, each addressing aspects of people, process, technology, and metrics:

- **Level 1: Manual & Unmeasured (Chaos)** – Highlights limited security awareness and collaboration, ad hoc processes, basic tools, and minimal metrics:

  - **People**: Limited communication about quality. Reactive approach.

  - **Process**: Manual quality checks. No integration into the development life cycle.

  - **Technology**: Disparate, non-integrated quality tools.

  - **Metrics**: Few, if any, quality metrics. Minimal tracking.

- **Level 2: Continuous Integration (CI)** – Some collaboration in security, initial integration of security checks during CI, basic automated tools, and basic metrics for vulnerability identification:

  - **People**: Some collaboration between development and QA teams on quality.

  - **Process**: Automated quality checks integrated into CI.

  - **Technology**: Basic automated testing tools integrated with CI.

  - **Metrics**: Basic quality metrics tracked during CI.

- **Level 3: Continuous Delivery (CD1)** – Enhanced team collaboration, full integration of security checks in CD processes, advanced assessment tools, and comprehensive security metrics:

  - **People**: Enhanced collaboration across all teams for quality assurance.

  - **Process**: Quality checks fully integrated into CD processes.

  - **Technology**: Advanced automated testing and quality tools for CD.

  - **Metrics**: Comprehensive quality metrics tracked, guiding CD decisions.

- **Level 4: Continuous Deployment (CD2)** – Strong focus on security in deployment, automated security processes, real-time monitoring tools, and detailed metrics for rapid response:

  - **People**: Strong focus on quality in deployment practices.

  - **Process**: Automated and streamlined quality processes for deployment.

  - **Technology**: Sophisticated tools for ensuring quality in deployments.

  - **Metrics**: Detailed quality metrics for deployment and feedback loops.

- **Level 5: Continuous Operations (CO)** – Organization-wide commitment to security, unified processes, continuous monitoring tools, and advanced metrics such as time to resolve security events:

  - **People**: Organization-wide commitment to quality in operations.

  - **Process**: Unified quality processes between development and operations.

  - **Technology**: Integrated tools for monitoring and maintaining quality.

  - **Metrics**: Advanced metrics, including SLOs and SLIs, to govern quality.

The continuous security CMM helps organizations assess and evolve their continuous security practices, ensuring security is seamlessly integrated across the software delivery life cycle. Each level defines specific focus areas for advancing security processes, technology, and metrics, guiding organizations toward a more proactive and responsive security posture. Organizations can determine their current maturity level by evaluating their practices against the criteria at each stage, thereby identifying areas for improvement and strategies for advancing to higher maturity levels.

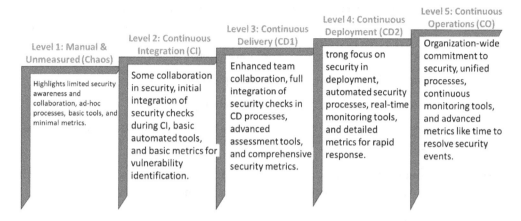

Figure 4.7 – Continuous security CMM

While there is no standard continuous security CMM, each organization should define one suitable for the organization. The model provided in *Figure 4.7* can be used as a basis for creating one specific to the organization.

Capability maturity levels – Continuous feedback

The model illustrated in *Figure 4.8* presents a continuous feedback CMM across five levels:

- **Level 1: Manual & Unmeasured** – Characterized by limited communication, ad hoc feedback collection, basic tools, and minimal metrics:

  - **People**: Limited communication and understanding of feedback's value.

  - **Process**: Ad hoc feedback collection, no formal process.

  - **Technology**: Basic tools, if any, for feedback collection.

  - **Metrics**: Few, if any, metrics to measure feedback implementation or impact.

- **Level 2: Continuous Integration** – Feedback is integrated within development phases, considered during CI, with initial tools for collection and basic metrics for frequency and impact:

  - **People**: Feedback is integrated within the development and integration phases.

  - **Process**: Feedback is considered during CI.

  - **Technology**: Initial tools for feedback collection integrated into CI.

  - **Metrics**: Basic metrics for feedback frequency and initial impact are used as part of CI process decisions.

- **Level 3: Continuous Delivery** – Enhanced collaboration, systematic feedback integration, advanced tools for analysis, and comprehensive metrics to assess implementation speed and impact:

  - **People**: Enhanced collaboration on feedback integration.

  - **Process**: Systematic feedback integration into CD processes.

  - **Technology**: Advanced tools for collecting and analyzing feedback.

  - **Metrics**: Comprehensive metrics to assess feedback implementation speed and impact used to determine delivery decisions.

- **Level 4: Continuous Deployment** – Strong focus on feedback in deployment, streamlined integration, real-time tools, and detailed metrics for deployment improvements:

  - **People**: Strong focus on feedback in deployment strategies.

  - **Process**: Streamlined feedback integration for rapid deployment adjustments.

  - **Technology**: Real-time feedback tools for deployment stages.

  - **Metrics**: Detailed metrics for feedback-driven deployment improvements.

- **Level 5: Continuous Operations** – Organization-wide commitment, unified feedback processes, integrated continuous tools, and advanced metrics, including system reliability improvements:

  - **People**: Organization-wide commitment to leveraging feedback.

  - **Process**: Unified feedback processes across development and operations.

  - **Technology**: Integrated tools for continuous feedback collection and analysis.

  - **Metrics**: Advanced metrics, including system reliability improvements due to feedback.

The model is useful for describing the maturity levels of an organization's continuous feedback practices, aligning them with broader goals of enhancing user satisfaction and reducing production failure rates. Each level defines specific areas for advancing feedback processes, technology, and metrics, guiding organizations toward more effective and efficient feedback integration. By evaluating this model, organizations can identify where they stand in terms of feedback maturity and plan strategic advancements.

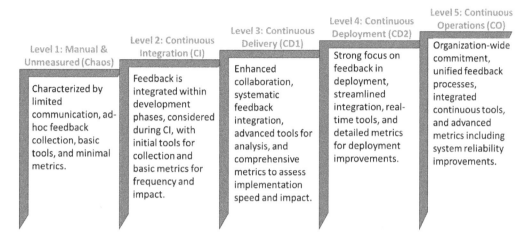

Figure 4.8 – Continuous feedback CMM

While there is no standard continuous feedback CMM, each organization should define one suitable for the organization. The model provided in *Figure 4.8* can be used as a basis for creating one specific to the organization.

Summary

This chapter provided a systematic and disciplined engineering approach to planning and implementing continuous testing, quality, security, and feedback. The chapter emphasized the necessity of an engineering approach in digital transformation, integrating technology with process and cultural changes. This included the Seven-Step Transformation Engineering Blueprint, explaining its application in continuous practices.

The integration of AI tools with the expertise of transformation consultants offers dynamic synergy in each step of the transformation process. AI contributes data-driven insights and automation to streamline decision-making, while consultants apply strategic oversight, industry knowledge, and critical interpretation of AI data to guide the transformation effectively. This collaborative approach ensures that organizational transformations are not only goal-aligned but also utilize cutting-edge technology for enhanced outcomes.

The chapter also detailed CMMs for continuous testing, quality, security, and feedback, demonstrating how these models guide transformations. The content is instrumental in understanding how to enhance practices to higher levels of capability maturity, ensuring a balanced, systematic, and incremental approach to transformation in the modern digital landscape.

In the next chapter, we will explore strategies for transformation goal setting and how to get organization alignment around the goals.

5

Determining Transformation Goals

This chapter explains a prescriptive methodology for determining goals for continuous testing, quality, security, and feedback transformations, to suit specific organizations, products, and services. Tools to help determine goals are described. The chapter builds on the engineering approach to digital transformations described in the last chapter. It also provides guidance on how to determine model applications for transformations.

By the end of this chapter, you will understand the importance of transformation goal alignment, how to determine a model application for transformations, and how to set goals for continuous testing, quality, security, and feedback.

This chapter is organized into the following sections:

- Transformation goal classifications
- The importance of transformation goals alignment
- Determining specific goals for a transformation
- Determining model applications
- Determining goals for continuous testing
- Determining goals for continuous quality
- Determining goals for continuous security
- Determining goals for continuous feedback

Let's get started!

> **Note**
>
> The scorecard examples discussed in this chapter are accessible as Excel files in the GitHub repository for this book.

Transformation goal classifications

For digital transformations to be successful, it is essential to establish goals in a way that supports clear direction, measurable outcomes, and alignment with the overall strategic vision and goals of the organization. Goals for digital transformations can be classified into the following seven categories:

- **Strategic goals**: Every transformation needs one or more overall strategic goals that focus on aligning digital transformation efforts with overarching organization visions and objectives to secure competitive advantage and market leadership. All the other goals must support the strategic goals.

    Example: Increase digital sales channels' revenue by 30% within the next 18 months to secure a top-three market position in our industry.

- **Agility goals**: Aim to enhance the organization's responsiveness to market changes and customer demands through the adoption of agile methodologies and flexible technology solutions.

    Example: Reduce product development cycle time by 40% by implementing agile development practices across all teams within 12 months.

- **Efficiency goals**: Target the improvement of operational processes and resource utilization to reduce costs and increase productivity through digital solutions.

    Example: Automate 50% of manual data entry processes within the next six months to achieve a 20% reduction in operational costs.

- **Stability goals**: Ensure the organization's digital infrastructure is robust and dependable, supporting continuous operations and minimizing downtime.

    Example: Achieve 99.9% uptime for all critical digital services over the next fiscal year through infrastructure enhancements and the implementation of a comprehensive disaster recovery plan.

- **Quality goals**: Focus on enhancing the quality of customer interactions, services, and products through the strategic use of digital technologies.

    Example: Improve customer satisfaction scores by 25% within one year by deploying a personalized customer service platform that leverages AI for enhanced customer interactions.

- **Security goals**: Concentrate on strengthening cybersecurity measures and ensuring compliance with relevant regulations to protect organizational and customer data.

Example: Eliminate all identified vulnerabilities in the organization's digital infrastructure within the next six months and achieve full compliance with industry security standards.

- **Team satisfaction goals**: Prioritize the enhancement of team engagement, satisfaction, and digital literacy through targeted training programs and the cultivation of a supportive digital culture.

    Example: Increase employee engagement scores by 20% within the next nine months by implementing a comprehensive digital literacy program and establishing a feedback-driven culture of continuous improvement.

Each of these goals must be SMART, that is, specific, measurable, achievable, relevant, and time-bound, providing a clear direction for digital transformation efforts. Focusing on these seven areas ensures organizations take a comprehensive approach to transformation that addresses both technical and human factors, driving success in the digital era. This not only supports a more structured and focused transformation effort but also helps align initiatives with the organization's broader strategic objectives.

Figure 5.1 illustrates the classifications of goals most useful for digital transformations. Each circle represents a different category of goals.

Figure 5.1 – Digital transformation goal classifications

These categories encompass the broad areas that organizations typically focus on during digital transformation efforts, highlighting the multifaceted approach required to achieve comprehensive digital change.

The importance of transformation goals alignment

Obtaining organizational and technical alignment of goals across the seven classifications of transformation goals is crucial for the following reasons:

- **Coherence in direction**: Alignment ensures all efforts contribute cohesively toward the broader vision and mission of the organization. Without alignment, efforts could be counterproductive or directionless, leading to a scattered and inefficient allocation of resources.

- **Synergy and efficiency**: When goals are aligned across teams, different initiatives can support and enhance each other, creating synergies that increase overall efficiency. Misalignment could result in duplicated efforts or initiatives that work at cross-purposes.

- **Risk mitigation**: Aligned goals across the organization help identify and manage risks comprehensively. If goals are misaligned, risks might be overlooked or inadequately managed, potentially leading to failures or breaches.

- **Optimized resource utilization**: Alignment ensures that the organization's resources are optimally utilized toward shared objectives. Misalignment can lead to a waste of time, money, and human capital.

- **Change management**: Aligned goals facilitate smoother change management as stakeholders understand the direction and purpose of transformation efforts. Misalignment can cause confusion and resistance among employees and other stakeholders.

- **Measurement and adaptation**: Aligned goals allow for the development of consistent metrics and KPIs, making it easier to track progress and make informed adjustments. Disparate goals make it challenging to measure success and to adapt strategies effectively.

- **Unified culture**: A common set of goals fosters a unified organizational culture that supports digital transformation. Misaligned goals can lead to a fragmented culture with varying degrees of acceptance and support for transformation efforts.

Figure 5.2 illustrates the concept of aligning transformation goals together with strategic goals.

Figure 5.2 – Transformation goals alignment

In my consulting work, I see many organizations that proceed to implementation without clarifying the goals, or at least some of the critical goals.

Negative consequences of misalignment in each classification

If we fail to align the goals, as depicted in *Figure 5.2*, it may lead to the following disadvantages:

- **Strategic goals**: If strategic goals are not aligned with other transformation goals, there could be a misallocation of IT resources, leading to projects that don't support the strategic aim of increasing market share or revenue, ultimately affecting competitive positioning.

- **Agility goals**: Without alignment, the organization may fail to respond quickly to market changes, resulting in missed opportunities and the inability to meet customer expectations, which can diminish market relevance.

- **Efficiency goals**: Misalignment here could lead to investments in technologies that do not integrate well with existing processes, reducing operational efficiency and failing to achieve the intended cost savings.

- **Stability goals**: If stability goals are not aligned, there could be a risk of system failures and outages, which can lead to significant operational disruptions and loss of customer trust.

- **Quality goals**: A lack of alignment may result in poor customer experiences due to inadequate or improper use of technology, which can lead to lower customer satisfaction and loyalty.

- **Security goals**: Misalignment between security goals and the technical infrastructure can lead to vulnerabilities and data breaches, potentially causing significant legal, financial, and reputational damage.

- **Team satisfaction goals**: Without alignment, teams may not receive the necessary support or resources to enhance their digital literacy and engagement, which can lead to low morale, high turnover rates, and a workforce that is ill prepared for a digital future.

In summary, alignment across all classifications of transformation goals ensures that the organization moves forward in a coordinated way, making the best use of resources, managing risks, and building a culture that supports sustained success in the digital era. Misalignment can lead to inefficiencies, increased risks, wasted resources, and failure to achieve the desired transformation outcomes.

Determining specific goals for a transformation

There are many goals within each category. If an organization has too many goals, it risks confusion across teams. Choosing a concise set of highest-priority specific goals for the transformation of an organization is critical. Determining an optimal set of goals involves a strategic and methodical approach. Here is a guide to navigating this process effectively.

Figure 5.3 shows a sequential process with four stages involved in setting and prioritizing goals for an organizational transformation.

Figure 5.3 – Transformation goal-setting process

The four stages are as follows:

1. **Transformation goals scorecard**: A scorecard method is used to assess and categorize transformation goals. The content and structure of the scorecard need to be arranged with the goal categories and goals in accordance with the strategic goals of the transformation. The transformation goals scorecard should be tailored around the strategic goals of the organization. The DevOps transformation goals scorecard shown in *Figure 5.4* is an example of a scorecard that I use as a baseline for DevOps transformations and is suitable to use as a baseline for scorecards specific to continuous testing, quality, security, and feedback.

    Within the scorecard, example goals are described for each of the goal classifications: agility, security, satisfaction, quality, efficiency, and stability. In preparation for *step 2*, a transformation team needs to be selected and informed about the workshop process.

2. **Transformation goals workshop**: The second stage is a collaborative approach, involving stakeholders in one or more workshops to understand and suggest scores for the goals in the scorecard. This workshop should include representation from leadership and relevant stakeholders from architecture development, QA, operations, tools, infrastructure, security functions, release management, project management, DevOps, and SRE functions. At this point, a time horizon should be set for the goals. A good practice is to set a time horizon such as six months. That is long enough to show significant progress, but not so long that the team that set the goals has moved on to other roles.

3. **Determine scores**: At the end of the workshop, the next step is to quantify or score the goals, based on criteria established in the scorecard and insights gathered during the workshop.

4. **Select highest-priority goals**: The final step in the process is to select the most critical goal(s) based on the ranking scores determined in the previous step, focusing efforts on the highest priorities for the transformation. Choose the top three ranked goals to focus on for the transformation at any time. If you choose too many, then the teams will be confused, resources will be spread too thin, and there will be an increased risk of failing the transformation's main goals.

This four-step goal-setting process is a cyclical process where the outcomes from the last step could loop back to influence the scorecard in future iterations. The overall flow is from conceptualization and categorization to collaboration, evaluation, and, finally, prioritization.

By following these steps, an organization can choose a concise set of high-priority, specific transformation goals that are aligned with its strategic vision, market position, stakeholder needs, and operational capabilities, setting the stage for a successful digital transformation.

Figure 5.4 is an example of a scorecard that I use for DevOps. The scorecard is designed to generate a rank number for priorities. The unit value defined for each goal must be selected to match the organization's state.

DevOps Transformation Goals Scorecard	Metric Unit	Importance	Current State	Desired State	Percent Improvement	Score (I x P%)	Rank
Agility		4.4			205%	9	1
Lead time: Duration from code commit until code is ready to be deployed to production.	# of days	5	5.0	2.0	250%	13	2
Release Cadence: Frequency of having releases ready for deployment to live production.	per month	5	0.2	0.6	300%	15	1
Fraction of Non-Value-Added Time: % of time employees are not spending on new features.	Fraction %	3	30%	20%	150%	5	15
Batch Size: Product teams break work into small batch increments.	1 to 10	5	3.0	5.0	167%	8	3
Visible Work: Workflow is visible throughout the pipeline.	1 to 10	4	5.0	8.0	160%	6	6
Security		2.7			233%	6	2
Security Events: # of times that serious business-impacting security events occur.	per year	2	2	1	400%	8	4
Unauthorized Access: # of times that unauthorized users access unauthorized information.	per year	3	1	1	100%	3	21
Fraction of timeemployees spend remediating security problems: % of time.	%	3	5%	3%	200%	6	9
Satisfaction		3.8			137%	5	5
Employee satisfaction with team: Employees recommend their team as a great to work with.	1 to 10	3	7.0	8.0	114%	3	20
Employee satisfaction with organization: Employees are likely to recommend their organization.	1 to 10	4	6.0	8.0	133%	5	13
Organization type: The culture has good communication flow, cooperation, and trust.	1 to 10	4	5.0	7.0	140%	6	11
Leader Style for Recognition: Leaders recognitize people and teams for outstanding work.	1 to 10	4	5.0	8.0	160%	6	6
Stability		4.0			152%	6	3
MTTR: Mean-Time-To-Recover (MTTR) from failure/service outage in production.	hours	4	1.0	0.7	154%	6	8
Code merges problems: % of time that code merges from development break the trunk branch.	%	4	15%	10%	150%	6	10
Quality		3.7			148%	5	4
Failures in Production: Frequently of failures requiring immediate action in live production.	per week	4	0.1	0.05	200%	8	4
Test and Data Available: Tests and test data are sufficient and readily available when needed.	1 to 10	3	7.0	9.0	129%	4	19
Customer Feedback: The organization seeks customer feedback and incorporates the feedback.	1 to 10	4	7.0	8.0	114%	5	14
Efficiency		3.3			143%	5	6
Unplanned work: % of time employees spend on all types of unplanned work, including rework.	%	3	15%	10%	150%	5	15
Operating Costs are Visible: Comprehensive metrics for **operating costs** are kept and visible.	1 to 10	3	5.0	7.0	140%	4	17
Capital costs are visable: Comprehensive metrics are kept for **capital** costs and visible.	1 to 10	3	5.0	7.0	140%	4	17
Backlog Visibility: Lean product management is practiced.	1 to 10	4	5.0	7.0	140%	6	11

Figure 5.4 – DevOps transformation goals scorecard

During the workshop, team members will choose an "importance score" value (I) from 1 to 5 to indicate how important each goal is toward achieving the strategic transformation goal. Then, the team determines what the current state and desired future state are. The desired percent (%) improvement between the current state and desired future state is computed as the value P. The overall score for each goal is then computed as I x P. Finally, the scores are ranked from 1 to n to determine the highest-priority goals.

Using AI chatbots to help determine transformation goals

AI chatbots can play a significant role in facilitating the setting of digital transformation goals in several innovative ways. Their capabilities can be leveraged to gather insights, automate tasks, and enhance decision-making processes. Here are some ways AI chatbots can help:

- **Data collection and analysis**:

    - **Surveys and feedback**: AI chatbots can conduct surveys and collect feedback from internal stakeholders and customers regarding pain points, needs, and expectations from digital transformation. This data is invaluable for setting relevant and impactful goals.

    - **Market insights**: Chatbots can be used to gather and analyze market trends and competitor information, helping organizations set goals that ensure they stay competitive and relevant.

- **Enhancing stakeholder engagement**:

    - **Stakeholder communication**: They can facilitate continuous communication with stakeholders by providing updates, collecting inputs, and answering queries about the digital transformation process. This keeps stakeholders engaged and allows for their expectations to be aligned with the transformation goals.

- **Facilitating collaboration**:

    - **Cross-functional collaboration**: By integrating with collaboration tools, AI chatbots can facilitate seamless communication across different departments and teams, ensuring that everyone is aligned and working toward the same digital transformation objectives.

- **Supporting decision-making**:

    - **Data-driven insights**: With access to vast amounts of data, AI chatbots can provide real-time insights and recommendations based on data analysis. This can help decision-makers set goals that are backed by data, improving the chances of success.

    - **Prioritization of goals**: Chatbots can assist in prioritizing digital transformation goals by analyzing the potential impact and feasibility of each goal, using criteria such as ROI, customer impact, and alignment with strategic objectives.

- **Tracking progress and feedback**:

  - **Monitoring progress**: They can track the progress of various initiatives against the set goals and provide regular updates to the relevant stakeholders, ensuring that any deviations are promptly addressed.

  - **Continuous feedback loop**: Chatbots can create a continuous feedback loop, collecting insights on what is working and what is not. This allows for the goals to be dynamically adjusted based on real-world performance and feedback.

- **Training and support**:

  - **Digital literacy**: AI chatbots can offer personalized learning experiences and support to employees, enhancing their digital skills and literacy. This supports a culture that is more adaptable to change, which is crucial for successful digital transformation.

  - **Change management**: Through interactive guides and FAQs, chatbots can address concerns and questions about digital transformation, helping to manage change resistance and foster a positive attitude toward new technologies and processes.

- **Innovation and experimentation**:

  - **Idea generation**: By analyzing trends and data, chatbots can suggest innovative ideas for digital products, services, or processes, encouraging a culture of experimentation and innovation.

In summary, AI chatbots can significantly streamline the process of setting and achieving digital transformation goals by automating and enhancing tasks related to data collection, stakeholder engagement, decision-making, and progress tracking. By leveraging the capabilities of AI chatbots, organizations can ensure that their digital transformation goals are well informed, strategically aligned, and adaptable to change.

Determining how many applications to transform at a time

The decision on whether an organization should orchestrate digital transformations one application at a time or tackle multiple applications simultaneously depends on several factors, including the organization's size, complexity, resource availability, strategic objectives, and risk tolerance. Both approaches have their advantages and disadvantages, and the optimal strategy often lies in a balanced, tailored approach. Here are some breakdowns to consider.

One application at a time

The following are the advantages of this approach:

- **Focused resources**: Concentrating on one application allows for a focused allocation of resources, including budget, personnel, and management attention.

- **Risk management**: Easier to manage risks as the scope is limited to one application, making it simpler to monitor and address issues as they arise.
- **Learning and adaptation**: Learnings from each transformation can be applied to subsequent projects, potentially improving efficiency and effectiveness over time.
- **Easier change management**: A smaller scope makes it easier to manage the change among users and stakeholders, facilitating smoother adoption and less disruption.

The following are the disadvantages:

- **Slower overall transformation**: This approach might extend the total time required to complete the digital transformation across the organization.
- **Delayed ROI**: The benefits of digital transformation may take longer to materialize organization-wide, as improvements are implemented in a sequential manner.

Multiple applications at a time

The following are the advantages of this approach:

- **Accelerated transformation**: Enables quicker realization of transformation goals across the organization, potentially leading to faster achievement of strategic objectives.
- **Simultaneous innovation**: Allows the organization to innovate and transform various aspects of the business at once, potentially unlocking synergies between different applications.
- **Competitive advantage**: Rapid transformation can provide a competitive edge in fast-moving markets by enabling quicker adaptation to market changes and customer needs.

The following are the disadvantages:

- **Resource strain**: Can place significant demand on resources, including personnel, budget, and management bandwidth, potentially leading to burnout and suboptimal execution.
- **Increased complexity and risk**: Managing multiple transformations simultaneously can significantly increase complexity and risk, including the risk of failure or disruptions to business operations.
- **Change management challenges**: The broader scope and faster pace of change can be more challenging for stakeholders to absorb, potentially leading to resistance or lower adoption rates.

Tailored approach

In practice, the best approach often involves a tailored strategy that considers the specific context and capabilities of the organization. Some considerations for a balanced approach include the following:

- **Strategic prioritization**: Start with applications that are most critical to achieving strategic objectives or those that offer the highest ROI.

- **Capability assessment**: Assess the organization's capacity to manage multiple projects, including the availability of skilled personnel and financial resources.

- **Risk appetite**: Consider the organization's tolerance for risk and potential disruptions to operations.

- **Incremental scaling**: Begin with a single application to build experience and capabilities, then gradually increase the scope of transformation efforts as confidence and competencies grow.

The tailored approach to digital transformation, as defined in the provided context, offers several advantages that can significantly benefit organizations looking to modernize their operations and enhance their competitive edge. Here are a few key advantages:

- **Strategic prioritization**: One of the primary advantages of the tailored approach is its emphasis on strategic prioritization. By focusing on applications that are most critical to achieving strategic objectives or those that offer the highest **return on investment** (**ROI**), organizations can ensure that their resources are allocated efficiently. This prioritization helps align the transformation efforts with the overall business goals, making it more likely that the initiatives will contribute directly to the organization's success.

- **Capability assessment**: Another advantage is the thorough capability assessment that the tailored approach advocates. This assessment evaluates the organization's capacity to manage multiple projects by considering the availability of skilled personnel and financial resources. By understanding its capabilities, an organization can tailor its transformation strategy to match its actual capacity, which reduces the risk of overextension and ensures that the transformation efforts are sustainable over the long term.

- **Risk appetite**: This is crucial because digital transformation can involve significant changes that might disrupt existing processes. As the tailored approach also considers the organization's tolerance for risk and potential operations disruptions, as well as integrating the organization's risk appetite into the transformation strategy, the approach helps in managing and mitigating potential risks effectively. This ensures that the organization does not undertake projects that could jeopardize its operational stability.

- **Incremental scaling**: Finally, the tailored approach advocates for incremental scaling, which is a significant advantage for managing change effectively. Starting with a single application allows the organization to build experience and capabilities gradually. This method helps in identifying potential issues and learning from them in a controlled environment before scaling up the efforts. Incremental scaling also helps in building confidence among stakeholders and reduces the resistance to change, as the benefits of transformation become evident gradually and reliably.

In summary, the tailored approach offers a strategic, risk-managed, capability-oriented, and incrementally scalable method for digital transformation. This approach not only aligns with the specific needs and conditions of the organization but also maximizes the chances of successful implementation and sustainable growth.

The choice between transforming one application at a time versus multiple applications should align with the organization's strategic goals and operational capabilities and the specific challenges and opportunities it faces in its digital transformation journey.

Model applications

Selecting an application to serve as a model for the transformations of other applications is an approach that offers several strategic advantages for an organization embarking on a digital transformation journey. This approach can effectively streamline the transformation process, reduce risks, and enhance the likelihood of success across the organization. Here are the key advantages:

- **Proof of concept**:

  - **Demonstrate value**: Using a single application as a model allows the organization to demonstrate the tangible benefits and value of digital transformation, which can help in securing buy-in from stakeholders across the board.

  - **Test and learn**: It provides an opportunity to test new technologies, processes, and strategies on a smaller scale before rolling them out organization-wide, allowing for adjustments and optimizations based on real-world feedback.

- **Risk management**:

  - **Controlled environment**: Transforming one application at a time allows for a more controlled environment to manage risks and address challenges without overwhelming resources or disrupting business operations significantly.

  - **Mitigate failures**: Lessons learned from any issues or failures can be applied to subsequent transformations, reducing the likelihood of similar problems occurring again.

- **Resource optimization**:

  - **Focused investment**: Resources can be more effectively allocated and managed when concentrated on a single application, ensuring that financial, human, and technical resources are used efficiently.

  - **Build expertise**: Concentrating on one application helps in building deep expertise and capabilities within the team, which can be leveraged in subsequent transformations.

- **Enhanced change management**:

  - **Smoother adoption**: A successful model application can serve as a case study to demonstrate the positive impact of transformation, thereby easing concerns and resistance among users and stakeholders.

  - **Cultural shift**: Early success can help foster a culture of innovation and openness to change, setting the stage for a more receptive environment for future transformations.

- **Streamlined implementation**:

  - **Standardization of processes**: The experience gained from transforming the model application can help in standardizing processes, methodologies, and best practices for digital transformation across the organization.

  - **Scalable strategies**: Insights from the model application can inform scalable strategies that can be adjusted and applied to other applications, potentially speeding up the transformation process.

- **Strategic alignment**:

  - **Alignment with business goals**: By carefully selecting an application that aligns closely with strategic business goals, organizations can ensure that the transformation delivers meaningful improvements to business performance and customer satisfaction.

  - **Feedback loop**: The initial transformation can serve as a feedback loop, providing valuable insights into how digital transformation impacts various aspects of the business and how it can be aligned with the overall business strategy.

By focusing on a single application as a model, organizations can create a blueprint for success that can be replicated across other applications, reducing the time, cost, and risk associated with broad-scale digital transformations. This approach not only facilitates a more manageable transformation process but also helps in building momentum and support for digital initiatives across the organization.

Determining model applications

It is critical to choose the model application carefully. This section describes the methodology that I use to help organizations choose an application.

Figure 5.5 depicts a detailed application transformation scorecard. The list that follows shows how the scorecard can be used to select a model application for transformations with the given criteria:

Rating scores: 0 = "Don't know/unsure", 1 = "Doesn't fit this characteristic", 2 = "Somewhat fits", 3 = "Mostly fits", 4 = "Good fit", 5 = "Perfect fit"	Rating (0, 1 to 5)
1. **Lead time**: The *application* will benefit from faster lead times, where lead time = time from backlog to deployment.	
2. **Leadership**: Leaders over this *application* are open to collaboration and will be sponsors of change.	
3. **Culture**: Team players that are associate with the *application* team (Product owners, Dev, QA, Ops, Infra, Sec, PM) are open to collaboration and change.	
4. **Application architecture**: The *application* is currently using, or planning to use, service-oriented, modular architectures.	
5. **Product Team size**: At least 15 people associated with the *application* team (includes Product owners, Dev, QA, Ops, Infra, Sec, PM) . Smaller projects may not show strong impact or justify substantive DevOps investment.	
6. **Duration**: The *application* is expected to be undergoing changes for more than a year. yield ROI.	
7. **Impact/Risk**: The *application* represents a good level of business impact and visibility but does not involve an extreme amount of risk for the business.	
8. **Frequent changes**: The *application* experiences frequent demands for changes from the business.	
9. **Tools**: The tools in the *application* toolchain will not all need to be replaced to implement a DevOps toolchain.	
10. **Effort per release**: Efforts to build, test and/or deploy releases of the *application* is significant and could be reduced significantly by automaton.	
Average	

Figure 5.5 – Model application transformation scorecard

Some applications can benefit from DevOps more readily than others. An average rating score of 3 or more is preferred.

1. **Define evaluation criteria**: The scorecard includes specific criteria for evaluating each application's suitability for a DevOps or DevSecOps transformation.

2. **Assign weightings**: Assign weightings to each criterion based on its importance to achieving the organization's DevOps transformation objectives. This could vary depending on the organization's specific goals, such as improving time to market, enhancing operational efficiency, or reducing risk.

3. **Rate each application**: Rate each application against these criteria on a scale provided (e.g., 0 for "don't know/unsure" to 5 for "very high"). The ratings should be based on current assessments, data, and stakeholder input.

4. **Calculate scores**: For each application, multiply the ratings by the weightings (if weightings are applied differently from a simple 1 to 5 scale) for each criterion and sum these values to get a total score. This score reflects the application's overall suitability for DevOps transformation.

5. **Compare and select**: Compare the total scores of all evaluated applications. The application with the highest score is the most suitable candidate for serving as a model for DevOps transformation. This application likely has the potential for significant improvements in lead times, supports a culture of innovation, and aligns with leadership objectives, making it an ideal candidate for demonstrating the benefits of DevOps practices across the organization.

This structured approach allows for an objective assessment of each application's potential for transformation, ensuring that resources are focused on areas with the highest impact and alignment with transformation goals.

Determining goals for continuous testing

Considering an organization that wants to transform toward continuous testing (continuous testing was defined in *Chapter 1*), examples of transformation goals for each of the six categories mentioned earlier in this chapter are described here:

- **Agility**:

  - We integrate automated testing into every phase of our CI/CD pipelines.

  - We adopt **test-driven development** (TDD) practices across our development teams.

  - We implement a feature flagging system that allows us to test new features in test and production environments with select user groups.

  - We increase the percentage of developed features that have associated automated tests before merging into the main branch, enhancing our test coverage.

- **Efficiency**:

  - We automate our regression testing.

  - We utilize ephemeral testing environments to scale testing efforts on demand, reducing environment provisioning time.

  - We optimize our test automation scripts to decrease execution times, thereby increasing testing frequency without additional resource costs over the next eight months.

  - We implement parallel testing strategies to reduce the overall testing cycle time.

- **Stability**:

  - We achieve a high percentage uptime of our continuous testing infrastructure, ensuring testing operations are uninterrupted.

  - We reduce production hotfixes through enhanced early detection of issues in the testing phase within the next year.

  - We ensure reduced downtime due to testing environment issues by implementing automated health checks and recovery processes.

  - We ensure low rollback rates for all deployments.

- **Quality**:

  - We increase the defect detection rate before code reaches the production environment, leveraging automated and manual testing synergies.

  - We improve user satisfaction scores by integrating continuous user experience testing into our release process.

  - Covering critical user journeys with automated tests reduces user-reported issues.

  - We conduct bi-weekly quality audits on our testing processes, aiming for improvement in our internal quality metrics quarter over quarter.

- **Security**:

  - We integrate automated security testing into every stage of our CI/CD pipelines.

  - We test for compliance with our security compliance requirements for our releases, reducing security-related incidents.

  - We conduct quarterly security training for our development and QA teams, aiming to reduce security breaches due to human error.

  - We implement continuous monitoring for security vulnerabilities in third-party dependencies, reducing the risk exposure time.

- **Team satisfaction**:

  - We establish a continuous learning culture, with the team receiving certification in automated testing tools within the next year.

  - We increase the satisfaction rate among our development and QA teams regarding the support and resources for continuous testing.

  - We increase team engagement scores by introducing a mentorship program pairing junior and senior professionals.

- We recognize and reward contributions to testing excellence monthly, leading to an increase in innovative testing practices adopted by the team.

These goals lay a foundation for a robust continuous testing strategy, emphasizing measurable improvements across agility, efficiency, stability, quality, security, and team satisfaction. *Figure 5.6* is an example of a continuous testing transformation goals scorecard suitable for the goal-setting workshop.

Continuous Testing Transformation Goals Scorecard	Metric Unit	Importance	Current State	Desired State	Percent Improvement	SCORE (I x P%)	RANK
Agility		4.0			201%	8	1
We integrate automated testing into every phase of our CI/CD pipelines.	% of phases	5	20.0%	90.0%	350%	18	1
We adopt Test-Driven Development (TDD) practices across our development teams.	% of teams	4	30.0%	80.0%	167%	7	4
We use feature flagging in test and production environments.	% of pipelines	4	25%	75%	200%	8	2
We automate tests for new features before merging a feature into the main branch.	% automated	3	35%	65%	86%	3	16
Efficiency		3.5			103%	4	3
We automate our regression testing.	% regression	4	50%	90%	80%	3	9
We utilize ephemeral testing environments to scale testing efforts on demand.	% environments	4	30.0%	70.0%	133%	5	7
We optimize our test automation scripts to decrease execution times.	% of scripts	3	25.0%	50.0%	100%	3	11
We implement parallel testing strategies to reduce the overall testing cycle time.	% test suites	3	25.0%	50.0%	100%	3	11
Stability		4.0			42%	2	5
We achieve high %uptime of our continuous testing infrastructure.	% uptime	4	60%	90%	50%	2	18
We reduce # of production hotfixes per release.	# of fixes	4	10	5	50%	2	19
We reduce downtime to testing environmentss using automated health checks and recovery.	# of hours	4	100%	65%	35%	1	20
We reduce the % of release rollbacks for all deployments.	% rollbacks	4	15%	10%	33%	1	21
Quality		3.8			45%	2	6
We use continuous testing automated and manual testing to reduce detects.	% /release	4	75%	95%	27%	1	23
We iintegrate continuous user experience testing into our release process.	1.0 to 10.0	4	7.0	8.5	21%	1	24
We automate testing of critical user journeys to reduce user-reported issues.	% journeys	4	50%	65%	30%	1	22
We conduct bi-weekly quality audits on our testing processes.	1.0 to 10.0	3	25%	50%	100%	3	11
Security		4.5			121%	5	2
We integrate automated security testing into every stage of our CI/CD pipelines.	% of stages	4	50%	80%	60%	2	17
We test for compliance with our security compliance requirements for our releases.	% compliance	5	45%	95%	111%	6	6
We conduct quarterly security training to reduce security breaches due to human error.	% of teams	5	35%	90%	157%	8	3
We scan for security vulnerabilities in third-party dependencies.	% pipelines	4	35%	90%	157%	6	5
Satisfaction		3.8			88%	3	4
We establish a continuous learning culture, and training in continuous testing.	% trained	3	30%	60%	100%	3	11
We increase the satisfaction score regarding the support and resources for continuous testing.	% satisfied	4	40%	70%	75%	3	15
We increase team engagement scores by introducing a mentorship program.	% engaged	4	45%	80%	78%	3	10
We recognize and reward contributions to testing excellence.	% of teams	4	30%	60%	100%	4	8

Figure 5.6 – Continuous testing goals scorecard example

In the example shown in *Figure 5.6*, the top three highest-ranked goals for continuous testing transformation are the following:

- We integrate automated testing into every phase of our CI/CD pipelines. From 20% to 90% of pipeline phases.

- We implement a feature flagging system that allows us to test new features in test and production environments with select user groups. From 25% to 75% of new features.

- We implement a feature flagging system that allows us to test new features in test and production environments with select user groups. From 35% to 90% of teams.

By resolving the goals to a short list in a workshop setting, the different team players will be more committed to the goals than if the goals were declared without their participation.

Determining goals for continuous quality

Considering an organization that wants to transform toward continuous quality (which was defined in *Chapter 1*), examples of transformation goals for each of the six categories defined earlier in this chapter are described here:

- **Agility**:

  - We complete all sprint retrospectives with actionable quality improvements, enhancing our sprint quality metrics.

  - We deploy new features based on user feedback quickly after receiving it, ensuring our product evolves in alignment with user needs.

  - We integrate automated testing in our CI/CD pipeline, achieving a reduction in manual testing time.

  - We implement feature flagging to test new features with select user groups, increasing our feature success rate.

- **Efficiency**:

  - We automate our regression testing, reducing the testing cycle time.

  - We reduce our overall software development life cycle time through process optimizations and tool integrations.

  - We achieve a year-over-year reduction in development costs by streamlining our development and deployment processes.

  - We enhance our resource utilization, decreasing idle time through better project management and forecasting.

- **Stability**:

  - We ensure a high percentage uptime for our production environment by implementing proactive monitoring and automated recovery processes.

  - We reduce critical production failures through comprehensive quality assurance practices and robust pre-deployment testing.

  - We reduce release rollbacks due to stability issues by adopting a phased release strategy and enhanced monitoring.

  - We establish a continuous monitoring framework that detects and resolves a high percentage of potential stability issues before they affect users.

- **Quality**:

  - We improve our customer satisfaction scores through targeted quality improvements and user experience enhancements.

  - We reduce user-reported issues with each new release by integrating user feedback into our development process.

  - We detect a high percentage of defects before production deployment, ensuring higher-quality releases.

  - We conduct monthly quality review meetings, resulting in improvement in our internal quality metrics quarter over quarter.

- **Security**:

  - We ensure a high percentage of our code base undergoes security code analysis before deployment, reducing security vulnerabilities.

  - We achieve compliance with all relevant security regulations, reducing compliance-related issues to 0.

  - We conduct quarterly security training for our development team, aiming to reduce security incidents caused by human error.

  - We implement a continuous security assessment process, identifying and remediating a high percentage of new vulnerabilities within 72 hours of detection.

- **Team satisfaction**:

  - We achieve high team satisfaction scores through enhanced communication, feedback mechanisms, and targeted training programs within the next year.

  - We provide monthly professional development opportunities for our team, aiming for a high percentage participation rate.

- We implement a peer recognition program that results in a high percentage of recognitions per team member per quarter, fostering a positive work culture.
- We ensure teams report their digital literacy as high through continuous learning and development programs.

Figure 5.7 is an example of a continuous quality transformation goals scorecard:

Continuous Quality Transformation Goals Scorecard	Metric Unit	Importance	Current State	Desired State	Percent Improvement	SCORE (I x P%)	RANK
Agility		3.5			903%	32	1
We complete all sprint retrospectives with actionable quality improvements.	% retrospecties	3	20.0%	60.0%	200%	6	5
We deploy new features based on user feedback quickly, ensuring our product evolves with user needs.	# days	4	35	14	60%	2	14
We integrate automated testing in our CI/CD pipeline, achieving a reduction in manual testing time.	% of pipelines	4	25%	75%	200%	8	2
Feature flagging helps test new features with select user groups, increasing our feature success	% of features	3	2%	65%	3150%	95	1
Efficiency		3.5			83%	3	3
We automate our regression testing, reducing the testing cycle time.	% regression automated	4	50%	90%	80%	3	9
We reduce our overall software development lifecycle time through process optimizations and tool integrations.	# days	4	20	10	50%	2	19
We achieve a year-over-year reduction in development costs by streamlining our development and deployment processes.	% of scripts	3	25.0%	50.0%	100%	3	10
We enhance our resource utilization, decreasing idle time through better project management and forecasting.	% of test suites	3	25.0%	50.0%	100%	3	10
Stability		4.0			65%	3	5
We ensure %uptime up-time for production environment using monitoring and automated recovery.	% uptime	4	60%	90%	50%	2	18
We use comprehensive quality assurance practices and robust pre-deployment testing.	failures / release	4	10	5	50%	2	19
We reduce release rollbacks due to stability issues using a phased release strategy and monitoring.	# rollbacks	4	5	2	60%	2	14
Test environment monitoring detects and resolves a high % of stability issues.	% environments	4	25%	50%	100%	4	7
Quality		3.5			52%	2	6
We improve our customer satisfaction scores through targeted user experience enhancements.	1 to 10	3	6.0	9.0	50%	2	23
We integrate user feedback into our development process.	# per release	4	20.0	7.0	65%	3	12
We detect a high % of defects before production deployment, ensuring higher quality releases.	% of defects	4	50%	65%	30%	1	24
We conduct monthly quality review meetings.	% monthly	3	40%	65%	63%	2	22
Security		4.5			121%	5	2
A high % of our codebase undergoes security code analysis before deployment.	% of codebase	4	50%	80%	60%	2	13
We achieve compliance with all relevant security regulations, reducing compliance-related issues.	% compliance	5	45%	95%	111%	6	6
We conduct quarterly security training for our development team.	% of teams	5	35%	90%	157%	8	3
Our continuous security assessment process, identifiesand remediateda high % of new vulnerabilities.	% of pipelines	4	35%	90%	157%	6	4
Satisfaction		3.5			76%	3	4
We achieve high team satisfaction scores through communication, feedback, and training programs.	1 to 10	4	6%	9%	50%	2	19
We provide monthly professional development opportunities for our teams.	% particiation	3	40%	70%	75%	2	17
Our peer recognition program that results in a high % of recognitions per team member.	% recognized	3	45%	80%	78%	2	16
Teams report their digital literacy high through continuous learning and development programs.	1 to 10	4	30%	60%	100%	4	7

Figure 5.7 – Continuous quality goals scorecard example

In the example shown in *Figure 5.7*, the top three highest-ranked goals for continuous quality transformation are as follows:

- We implement feature flagging to test new features with select user groups, increasing our feature success rate.

- We integrate automated testing in our CI/CD pipeline, achieving a reduction in manual testing time.

- We conduct quarterly security training for our development team, aiming to reduce security incidents caused by human error.

By resolving the goals to a short list in a workshop setting, the different team players will be more committed to the goals than if the goals were declared without their participation.

Determining goals for continuous security

Considering an organization that wants to transform toward continuous security (which was defined in *Chapter 1*), examples of transformation goals for each of the six categories defined earlier in this chapter are described here:

- **Agility**:

  - We integrate security assessments into sprints, ensuring a high percentage of new code undergoes security review before release.

  - We deploy security patches quickly after identification, maintaining our commitment to rapid response across all systems.

  - We adapt our security protocols based on emerging threats quickly after discovery, staying ahead of potential vulnerabilities.

  - We conduct agile retrospectives frequently, focused on security practices, continuously refining our approach based on the latest insights.

- **Efficiency**:

  - We automate a high percentage of our routine security monitoring tasks, reducing manual effort and increasing our operational efficiency.

  - We decrease the time to detect security threats through enhanced analytics and machine learning tools.

  - We cut down the resolution time for non-critical security alerts, optimizing our security operations center's effectiveness.

  - We streamline our incident response process, achieving improvement in the speed of containment and remediation actions.

- **Stability**:

  - We achieve a high percentage of uptime for our critical security systems through robust design and redundant configurations.

  - We ensure zero security-induced downtimes by implementing fail-safe mechanisms in our deployment processes.

  - We maintain operational stability, with a low percentage of security updates causing disruptions to system performance.

  - We ensure continuous operation of security monitoring tools, with low levels of unplanned downtime.

- **Quality**:

  - We maintain a high user satisfaction score regarding the security and privacy features of our services.

  - We ensure all customer-facing applications exhibit a low number of high-risk vulnerabilities at the time of release.

  - We achieve a high confidence score from customers in our data protection practices through transparent communication and robust security measures.

  - We reduce the number of security-related customer complaints through proactive engagement and education efforts.

- **Security**:

  - We identify and remediate a high percentage of critical vulnerabilities within 48 hours of detection.

  - We ensure a high percentage of compliance with international cybersecurity standards across all operations.

  - We reduce the frequency of security incidents through continuous improvement in our security posture.

  - We achieve a year-over-year reduction in the average time to resolve security events, demonstrating our increasing efficiency in incident management.

- **Team satisfaction**:

  - We increase the security team's job satisfaction score through targeted development programs and a supportive work environment.

  - We ensure development teams receive up-to-date training on secure coding practices annually.

- We foster a culture of security awareness, achieving a 90% pass rate on security knowledge assessments across all departments.

- We implement a recognition program for security best practices, resulting in at least one recognized contribution from each department per quarter.

Figure 5.8 is an example of a continuous security transformation goals scorecard:

Continuous Security Transformation Goals Scorecard	Metric Unit	Importance	Current State	Desired State	Percent Improvement	SCORE (I x P%)	RANK
Agility		3.5			58%	2	3
We integrate security assessments into sprints - new code undergoes security review before release.	% of sprints	3	50.0%	80.0%	60%	2	19
We deploy security patches quickly after identification.	# of days	4	35	14	60%	2	8
We adapt our security protocols based on emerging threats, to reduce potential vulnerabilities.	# of days	4	60	30	50%	2	12
We conduct agile retrospectives frequently, focused on security practices.	% retrospectives	3	40%	65%	63%	2	18
Efficiency		3.5			45%	2	6
We automate a high % of security monitoring tasks, reduce manual effort and increase efficiency.	% automated	4	50%	90%	80%	3	2
We limit the time to detect security threats through enhanced analytics and machine learning.	# days	4	20	10	50%	2	12
We reduce security alerts optimizing our security operations center.	# days	3	40	30	25%	1	23
We streamline our incident response process, limiting the time for containment and remediating	# days	3	40	30	25%	1	23
Stability		3.8			46%	2	5
We achieve a high uptime for security systems through robust design and redundant configurations.	% uptime	4	75%	90%	20%	1	22
We implement fail-safe mechanisms in deployment processes.	% of pipelines	4	40%	65%	63%	3	5
We maintain operational stability, with a low % of security updates causing disruptions.	% disruptions	3	40%	20%	50%	2	20
We ensure continuous operation of security monitoring tools.	% uptime	4	60%	90%	50%	2	11
Quality		4.0			48%	2	4
We maintain a high user satisfaction score regarding the security and privacy features.	1 to 10	4	6.0	9.0	50%	2	12
We ensure a low number of high-risk vulnerabilities at the time of release.	# per release	4	20.0	7.0	65%	3	4
We achieve a high confidence score from customers in our data protection practices through transparent communication and robust security measures.	1 to 10	4	7.0	9.0	29%	1	21
We reduce the number of security-related customer complaints through ngagement and education.	# per release	4	20	10	50%	2	12
Security		4.5			65%	3	1
We identify and remediate a high % of critical vulnerabilities within 48 hours of detection.	% compliance	4	50%	80%	60%	2	6
We ensure a high % compliance with international cybersecurity standards across all operations.	% compliance	5	45%	95%	111%	6	1
We practice continuous improvement in our security posture.	incidents/release	5	75%	45%	40%	2	17
We reduce average time to resolve security events.	days to resolve	4	20	10	50%	2	12
Satisfaction		3.3			78%	3	2
We increase the security team's job satisfaction with development programs and a support.	1 to 10	4	5.0	9.0	80%	3	2
We ensure development teams receive up-to-date training on secure coding practices annually.	% particiation	3	40%	70%	75%	2	10
We achieve a 90% pass rate on security knowledge assessments across all departments.	% departments	3	45%	80%	78%	2	9
We recognize security security contributions from each department per quarter.	% departments	3	50%	90%	80%	2	6

Figure 5.8 – Continuous security goals scorecard example

In the example shown in *Figure 5.8*, the top three highest-ranked goals for continuous security transformation are as follows:

- We ensure a high percentage of compliance with international cybersecurity standards across all operations.

- We automate a high percentage of our routine security monitoring tasks, reducing manual effort and increasing our operational efficiency.

- We increase the security team's job satisfaction score through targeted development programs and a supportive work environment.

By resolving the goals to a short list in a workshop setting, the different team players will be more committed to the goals than if the goals were declared without their participation.

Determining goals for continuous feedback

Considering an organization that wants to transform toward continuous feedback (which was defined in *Chapter 10*, examples of transformation goals for each of the six categories described earlier in this chapter are described here:

- **Agility**:

  - We establish a real-time feedback loop with our users, reducing the cycle time from feedback receipt to feature enhancement.

  - We integrate customer feedback directly into our sprint-planning process, ensuring that each sprint's work is directly influenced by user insights.

  - We deploy a user acceptance testing platform for new features, allowing us to gather user feedback and iterate improvements quickly after deployment of the initial release.

  - We implement an automated system for collecting, analyzing, and prioritizing feedback.

- **Efficiency**:

  - We automate the feedback analysis process, reducing the time required to categorize and act on feedback.

  - We reduce operational costs by streamlining feedback-driven development processes, eliminating inefficiencies in how user insights are integrated.

  - We enhance our product analytics tools to provide actionable insights from user behavior, increasing the speed of feedback implementation.

  - We implement a continuous delivery pipeline that automatically updates based on user feedback scores, optimizing resource use and reducing manual deployment efforts.

- **Stability**:

  - We ensure a high percentage uptime for our user feedback collection and analysis systems, supporting uninterrupted feedback loops.

  - We maintain system stability during peak feedback periods by scaling our infrastructure automatically, ensuring no downtime during critical release feedback phases.

  - We reduce the incidence of feedback-related system failures through robust testing and monitoring practices, enhancing overall system reliability.

  - We implement failover mechanisms for all critical feedback processing components, guaranteeing continuous operation with no single point of failure.

- **Quality**:

  - We improve overall product quality by incorporating feedback into development for a high performance of our pipeline stages , thus aiming for an increase in customer satisfaction scores.

  - We ensure all new features meet a high percentage positive feedback rate before wide release, demonstrating our commitment to quality based on user insights.

  - We decrease the number of post-release quality issues by integrating continuous user testing throughout the development process.

  - We establish a quality benchmark that includes user feedback metrics, aiming to improve these benchmarks every quarter.

- **Security**:

  - We integrate user feedback into our security review process, aiming to identify and remediate a high percentage of security concerns highlighted by the security organization and users quickly.

  - We ensure all user feedback channels are secure and compliant with data protection regulations, protecting user data.

  - We decrease user-reported security incidents by implementing continuous security training for our team, informed by common themes in user feedback.

  - We conduct regular security audits informed by user feedback trends, aiming to strengthen our cybersecurity measures continuously.

- **Team satisfaction**:

  - We increase team engagement and satisfaction by incorporating team feedback into operational improvements, measured through quarterly surveys.

  - We enhance digital literacy across the organization, driven by a feedback-informed training curriculum.

- We establish a team feedback portal that leads to process changes, directly enhancing team satisfaction and productivity.
- We recognize and reward contributions to the feedback process, aiming to increase team morale and participation in continuous feedback initiatives.

Figure 5.9 is an example of a continuous feedback transformation goals scorecard:

Continuous Feedback Transformation Goals Scorecard	Metric Unit	Importance	Current State	Desired State	Percent Improvement	SCORE (I x P%)	RANK
Agility		3.8			59%	2	3
We establish a real-time feedback loop with our users.	# of days	4	35	14	60%	2	11
We integrate customer feedback directly into our sprint planning process.	% of sprints	4	40%	65%	63%	3	6
Our user acceptance testing platform for new features, allows us to gather user feedback.	% Acceptance Tested	4	50%	90%	80%	3	3
We implement an unified automated system for collecting, analyzing, and prioritizing feedback.	% of applications	3	60%	80%	33%	1	21
Efficiency		3.5			45%	2	6
We automate feedback analysis, reducing the time required to categorize and act on feedback.	% automated	4	50%	90%	80%	3	3
We streamline feedback processes, to reduce inefficiencies in how user insights are integrated.	# of days	4	40	30	25%	1	22
Our product analytics tools provide quick actionable insights from user behavior.	# of days	3	40	30	25%	1	24
Our continuous delivery pipelines automatically update based on user feedback scores.	# of pipelines	3	40	60	50%	2	18
Stability		3.8			46%	2	5
We ensure high % uptime for our user feedback collection and analysis systems.	% uptime	4	75%	90%	20%	1	23
We scale our systems automatically, ensuring uptime during critical release feedback phases.	% of pipelines	4	40%	65%	63%	3	6
We use robust testing and monitoring practices, enhancing overall system reliability.	failures per month	3	40	20	50%	2	18
We implement failover mechanisms for all critical feedback processing components.	% of pipelines	4	60%	90%	50%	2	14
Quality		4.0			47%	2	4
We incorporate feedback to development from a high % our pipeline stages.	1 to 10	4	6.0	9.0	50%	2	15
We ensure all new features meet a high % positive feedback rate before wide release.	1 to 10	4	7.0	9.0	29%	1	20
We integrate continuous user testing throughout the development process.	% of pipelines	4	50%	80%	60%	2	9
Our quality benchmark includes user feedback metrics.	1 to 10	4	6.0	9.0	50%	2	15
Security		4.5			78%	4	1
We integrate user feedback into our security review process.	#days	4	10.0	5.0	50%	2	15
User feedback channels are secure and compliant with data protection regulations.	% of compliance	5	45%	95%	111%	6	1
Our continuous security training for our teams, is informed by common themes in user feedback.	incidents per release	5	10.0	5.0	50%	3	6
We conduct regular security audits informed by user feedback trends.	audits per month	4	5	10	100%	4	2
Satisfaction		3.3			78%	3	2
We incorporate team feedback into operational improvements, measured through surveys.	1 to 10	4	5.0	9.0	80%	3	3
We enhance digital literacy across the organization, driven by feedback-informed training.	% particiation	3	40%	70%	75%	2	13
Our team feedback portal leads to process changes, enhancing satisfaction and productivity.	% departments	3	45%	80%	78%	2	12
We recognize contributions to the feedback process.	% of departments	3	50%	90%	80%	2	9

Figure 5.9 – Continuous feedback goals scorecard example

In the example shown in *Figure 5.9*, the top three highest-ranked goals for continuous feedback transformation are as follows:

- We ensure all user feedback channels are secure and compliant with data protection regulations, protecting user data.

- We ensure all user feedback channels are secure and compliant with data protection regulations, protecting user data.

- We deploy a user acceptance testing platform for new features, allowing us to gather user feedback and iterate improvements quickly after deployment of the initial release.

By resolving the goals to a short list in a workshop setting, the different team players will be more committed to the goals than if the goals were declared without their participation.

Summary

This chapter serves as a cornerstone for anyone looking to navigate the tumultuous waters of change in today's fast-paced world. It meticulously outlined the significance of aligning organizational aspirations with actionable and measurable goals across various domains, including agility, efficiency, stability, quality, security, and team satisfaction. This narrative encourages leaders to think strategically about the future they wish to create and the steps necessary to bring that vision to fruition.

The chapter broke down complex concepts into relatable and actionable insights, making it an invaluable resource for decision-makers at all levels. Through a series of examples and best practices, it demystified the process of goal setting in a transformation context, emphasizing the critical role of continuous feedback and adaptation. This approach fosters a culture of innovation and continuous improvement and ensures that transformation efforts are grounded in the reality of operational capabilities and market demands. The focus on SMART goals acts as a beacon for organizations, guiding them toward outcomes that are both ambitious and achievable.

This chapter was a call to action for organizations seeking to embrace change proactively. It highlighted the transformative power of well-defined goals and the positive ripple effects they can have on all aspects of an organization. From enhancing customer satisfaction to bolstering security measures and fostering a supportive culture among teams, the chapter outlined a holistic framework for sustainable growth and resilience. It is not just about encouraging adherence to best practices but about inspiring a fundamental shift in how organizations perceive and approach change. This chapter empowers and equips leaders to turn the challenges of today into the successes of tomorrow.

The next chapter will explain how organizations can conduct current state discovery for the transformation.

6

Discovery and Benchmarking

This chapter explains the methodology and tools for discovering the current state of an organization's people, processes, and technologies relevant to the transformation of mastering continuous testing, quality, security, and feedback. This is necessary to formulate a benchmark starting point and priorities for transformations. The methodologies and tools described in this chapter include surveys, interviews, workshops, **gap assessments**, and **value stream mapping**.

By the end of this chapter, you will understand a methodology for discovery and benchmarks and the discovery tools for gap assessments and value stream mapping. We'll conclude the chapter by discussing how to apply discovery and benchmarks for continuous testing, quality, security, and feedback.

This chapter is organized under the following headings:

- Methodology for discovery and benchmarks
- Understanding current-state discovery
- Understanding gap assessments
- Understanding CSVSM
- How generative AI can be used to accelerate discovery and benchmarking

Let's get started!

Technical requirements

Executable versions of tools referenced in this book are available on the resources page that is associated with this book: https://github.com/PacktPublishing/Continuous-Testing-Quality-Security-and-Feedback.

Methodology for discovery and benchmarks

Chapter 4 outlined the Seven-Step Transformation Engineering Blueprint that I recommend for guiding digital transformations. *Step 4* of the transformation, called *Assessment*, is the step that includes discovery and benchmarking. There are four components for conducting discovery and benchmarking, as shown in *Figure 6.1*.

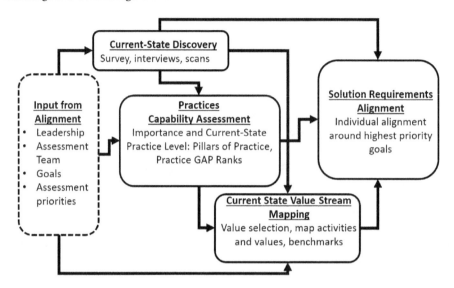

Figure 6.1 – Methodology for discovery and benchmarks

The discovery and benchmarking methodologies are the same as I described for DevOps in my book titled *Engineering DevOps*.

The four components are as follows:

- **Current state discovery**: Using the output of the goals determined in the prior steps in the seven-step transformation process, conduct surveys, interviews, and scans of the current environment and practices for continuous testing, quality, security, and feedback.

- **Practices capability assessment**: Using the output of the *current state discovery* component, determine the importance and current state of practices, and gaps for each practice in the pillars of practice for continuous testing, quality, security, and feedback.

- **CSVSM**: Using the output of the *practices capability assessment* component, select values, map the current state, and determine benchmarks for the current state of value streams.

- **Solutions requirements analysis:** Using the output of the *current state discovery, practices capability assessment*, and *CSVSM* components, determine the priorities for aligning the solution to the goals for the organization's transformation.

The discover methodology described by the preceding steps ensures that the organization thoroughly considers relevant data and current state.

Understanding current state discovery

Using surveys and interviews during the *assessment* phase of a digital transformation is vital for gathering comprehensive, nuanced insights that inform strategy, engage stakeholders, manage risks, and guide continuous improvement. These tools help ensure that the transformation efforts are well grounded in the organization's realities, increasing the chances of achieving meaningful and sustainable change.

Surveys

Surveys are a powerful tool to discover the current state of an organization's people, processes, and technologies as part of an overall assessment approach. They complement interviews and other data collection methods by providing quantitative insights that can be analyzed to identify trends, patterns, and areas for improvement.

Why conduct surveys?

Here's why surveys are important and how they should be conducted:

- **Broad reach and scalability**: Surveys can reach a large number of respondents across various levels and departments within an organization, providing a comprehensive view of the current state.

- **Quantifiable data**: They produce quantifiable data that can be easily analyzed to identify trends, compare subgroups, and measure against benchmarks or over time.

- **Anonymity and honesty**: Surveys can offer anonymity, which might encourage more honest responses, especially on sensitive topics.

- **Standardization**: Surveys ensure that the same information is collected from all respondents, making it easier to compare and analyze responses.

- **Efficiency**: They can be more time and cost effective than conducting individual interviews, especially for larger organizations.

How to conduct surveys

The steps are defined as follows:

1. **Define objectives**: Clearly define what you want to learn from the survey. This will guide the development of your questions and the analysis of the data.

2. **Design the survey**: Develop questions that are clear, concise, and directly related to your objectives. Ensure questions are unbiased and do not lead respondents toward a particular answer.

3. **Choose the right platform**: Select a survey platform that is accessible to all participants and provides the features you need (e.g., anonymity options, data analysis tools).

4. **Distribute the survey**: Send the survey to a carefully selected sample or the entire population of interest. Ensure to communicate the purpose of the survey, how the data will be used, and any incentives for participation.

5. **Maximize response rates**: Use reminders and follow-ups to encourage participation. Consider timing and make sure the survey is neither too long nor too complex.

6. **Analyze the data**: Use statistical methods to analyze the data. Look for significant trends, correlations, and differences between groups.

7. **Report findings**: Present the findings in a clear, actionable manner. Highlight key insights, potential areas for improvement, and recommendations based on the data.

8. **Take action and follow up**: Use the survey findings to inform strategic decisions and improvement initiatives. It's also crucial to communicate back to participants what changes or actions have been taken because of their feedback.

Best practices

Here are a few things we can implement to ensure we get the best out of this process:

-   **Ensure anonymity and confidentiality**: Make it clear how you will protect respondents' privacy and how the data will be used. This encourages participation and honesty.

-   **Keep it relevant**: Ensure every question contributes to your objectives. Unnecessary questions can reduce completion rates and dilute valuable insights.

-   **Be mindful of bias**: Avoid leading questions or wording that could bias responses. Neutral language helps in getting more accurate data.

-   **Consider timing and frequency**: Choose an appropriate time to distribute the survey and avoid over-surveying, which can lead to survey fatigue.

Let's understand this with the help of an example.

Example survey

Figure 6.2 shows an example survey for continuous testing practices. For each practice statement in the survey, the user enters an importance score, a current level of practice score, and any clarification comments.

Survey	(I) Importance	(P) Practice
Test Automation		
Unit tests are automated for every new piece of code.		
Integration tests are developed to ensure component interactions work as intended.		
System tests are automated to verify the application behaves as expected.		
User acceptance tests are automated to confirm the system meets user requirements.		
Test suites are continuously updated and refactored for relevance and efficiency.		
API testing is included in the automation strategy for interface reliability.		
Performance tests are automated and regularly executed to monitor system responsiveness.		
Integration with Development		
Test-Driven Development (TDD) is practiced, with tests written before code.		
Behavior-Driven Development (BDD) is employed to ensure alignment with expected		
Automated testing is integrated into each stage of the CI/CD pipeline for immediate feedback.		
Pair programming is encouraged to enhance code quality and facilitate instant testing.		
Feature toggles are used for managing and testing new features.		
Security testing is integrated early in the development cycle.		
Static code analysis tools are utilized during development to catch potential issues early.		
Feedback		
Dashboards are set up for immediate visibility of test results.		
Alerts for test failures are configured for quick response.		
Trends in test results are analyzed to identify areas for improvement.		
A feedback loop for test strategy enhancement is established.		
Test logs and details are made easily accessible for swift issue diagnostics.		
Visual testing tools are employed for detecting UI changes.		
AI and machine learning are leveraged for predictive analysis of test outcomes.		

Figure 6.2 – Example survey for continuous testing

Conducting surveys is an effective way to gather valuable data about an organization's people, processes, and technologies. When done correctly, surveys can provide the insights needed to drive strategic improvements and achieve organizational goals.

Interviews

Conducting interviews to discover the current state of an organization's people, processes, and technologies is a critical part of an overall assessment approach. These interviews help in understanding the existing conditions, identifying gaps, and uncovering opportunities for improvement.

Why conduct interviews?

Here's why interviews are important and how they should be conducted:

- **Capture qualitative insights**: Interviews provide deep, qualitative insights into the experiences, perceptions, and opinions of the organization's members regarding its people, processes, and technologies.

- **Identify strengths and weaknesses**: Through conversations, strengths and weaknesses within the organization's current setup can be identified, which might not be evident through quantitative methods such as surveys.

- **Facilitate stakeholder engagement**: Involve stakeholders directly, making them feel valued and heard, which can increase buy-in for any change initiatives that result from the assessment.

- **Developing customized solutions**: Understanding specific challenges and contexts allows for the development of tailored solutions that are more likely to succeed.

- **Uncover hidden issues**: Interviews can reveal issues and bottlenecks that are not documented or are under-recognized within formal processes.

How to conduct interviews

Figure 6.3 illustrates the suggested steps for conducting interviews.

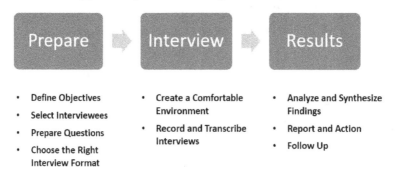

Figure 6.3 – Example interview process

The steps are defined as follows:

1. **Define objectives**: Clearly outline what you aim to learn from the interviews regarding the organization's people, processes, and technologies.

2. **Select interviewees**: Choose a diverse range of participants from different levels and departments within the organization to get a comprehensive view.

3. **Prepare questions**: Design open-ended questions that encourage detailed responses. Ensure questions cover all areas of interest but remain flexible to explore new insights that emerge during interviews.

4. **Choose the right interview format**: Decide whether interviews will be conducted in person, over the phone, or via video conference based on the availability of participants and the need for visual cues or demonstrations.

5. **Create a comfortable environment**: Make sure interviewees feel comfortable and understand that the purpose of the interview is to improve, not to judge or blame.

6. **Record and transcribe interviews**: With consent, record interviews for accuracy in capturing information and transcribe them for analysis.

7. **Analyze and synthesize findings**: Look for patterns, trends, and discrepancies in the responses. Identify common issues, unique insights, and potential areas for improvement.

8. **Report and action**: Compile a report summarizing the findings, insights, and recommendations. Use this report as a basis for planning and implementing changes.

9. **Follow up**: After implementing changes, follow up with participants to assess the impact and effectiveness of those changes. This can also help in maintaining engagement and showing appreciation for their input.

Best practices

Here are a few things we can implement to ensure we get the best out of this process:

- **Ensure confidentiality**: Assure participants that their responses will be kept confidential. This encourages openness and honesty.

- **Be objective and non-judgmental**: Approach each interview with an open mind and avoid making judgments based on responses.

- **Use skilled interviewers**: Interviewers should have strong interpersonal skills, be good listeners, and be able to probe deeper without leading the interviewee.

Conducting interviews as part of an overall assessment approach is invaluable for gaining a deep understanding of an organization's current state. It enables the identification of actionable insights that can guide strategic planning and operational improvements.

Example interview questions

The following questions are examples designed to uncover insights into existing processes, tools, challenges, and the team's readiness for a transformation to continuous testing. Similar questions can be applied for continuous quality, security, and feedback:

- Can you describe the current software testing process in your team/organization?

  - *Purpose*: This open-ended question aims to gather comprehensive information on the existing testing procedures, including manual and automated testing practices, the integration of testing in the development life cycle, and the roles involved in the testing process.

- What tools and technologies are currently in use for testing, and how effective do you find them?

  - *Purpose*: This question seeks to understand the toolset in place, including any test automation tools, test management software, and development environments. It also opens a discussion on the perceived effectiveness of these tools, potential gaps, and any limitations encountered.

- How do you currently handle test data management and test environment configuration?

  - *Purpose*: Test data management and the configuration of test environments are critical components of continuous testing. This question aims to uncover practices around the creation, maintenance, and use of test data and environments, identifying areas for improvement or automation.

- Can you share any challenges or bottlenecks you've encountered with your current testing process?

  - *Purpose*: Identifying existing challenges and bottlenecks is essential for planning a successful transition to continuous testing. This question encourages interviewees to discuss obstacles in efficiency, quality, coverage, or collaboration within the testing process.

- In your opinion, what are the key areas for improvement in our testing processes to support a shift toward continuous testing?

  - *Purpose*: This question directly engages the interviewee in identifying potential areas for improvement and aligning the goals of the transition to continuous testing. It can reveal insights into the readiness of the team, skill gaps, and the need for new tools or processes.

These questions are designed to facilitate a comprehensive understanding of the current testing landscape, the challenges faced, and the opportunities for enhancing testing practices through the transition to continuous testing. The responses will provide valuable insights for developing a tailored, effective strategy for implementing continuous testing within the organization.

Understanding gap assessments

Gap assessments play a critical role in the context of digital transformations, acting as a strategic tool to compare the current state of an organization's processes, technologies, and capabilities against its desired future state or industry benchmarks and know good practices. The primary goal of a gap assessment is to identify the differences (gaps) between where an organization is and where it wants or needs to be to achieve its digital transformation objectives. These assessments are vital for several reasons, and their conduct and the utilization of their results follow a structured approach.

Figure 6.4 shows an example gap assessment summary. A **gap score** is calculated for the practice area according to importance and current practice level. The set of scores across all practice areas is ranked to determine the practice area priorities.

Continuous Testing (CT) Practices	(I) Importance	(P) Practice Level	(G) GAP	# Practices	Rank
Test Automation	4.1	2.1	7.6	7	3
Integration with Development	3.0	1.6	6.9	7	5
Feedback	4.0	2.9	5.8	7	7
Testing Metrics	4.0	3.0	7.8	7	2
Risk-Based Testing	4.3	2.3	7.4	7	4
Test Environment and Data Management	4.0	2.0	8.0	7	1
Collaboration and Communication	4.0	3.7	5.3	7	8
Continuous Learning and Adaptation	4.0	2.6	6.7	7	6
Overall Assessment	3.9	2.5	6.9	56	
	Importance Average	Practice Level Average	Average GAP Level		

Figure 6.4 – Example gap assessment

Why gap assessments are important

Here's why gap assessments are important and how they should be conducted:

- **Identify shortcomings and opportunities**: Gap assessments help in pinpointing specific areas where the organization falls short of its digital transformation goals or industry standards, offering a clear picture of areas for improvement or investment.

- **Prioritize initiatives**: By highlighting the most significant gaps, organizations can prioritize the initiatives that will have the greatest impact on achieving their digital transformation objectives, ensuring efficient allocation of resources.

- **Customized strategy development**: The insights gained from gap assessments allow organizations to develop tailored strategies that address their unique challenges and leverage their specific strengths, rather than adopting a one-size-fits-all approach.

- **Risk management**: Understanding the gaps helps in identifying potential risks and obstacles to the transformation effort, allowing for the development of mitigation strategies.

- **Measure progress**: By establishing a baseline of the current state, gap assessments enable organizations to measure progress over time, providing a quantitative way to track the effectiveness of their digital transformation efforts.

How gap assessments are conducted

The steps are defined as follows:

1. **Define objectives and scope**: Clearly outline the objectives of the digital transformation and define the scope of the gap assessment, including the processes, technologies, and capabilities to be evaluated.

2. **Assess current state**: Collect data on the current state of the organization's technology, processes, and people. This can be done through surveys, interviews, workshops, and review of existing documentation.

3. **Define desired future state**: Articulate the vision for the organization's future state post-digital transformation, including the adoption of new technologies, process improvements, and capability enhancements.

4. **Identify gaps**: Compare the current state with the desired future state to identify gaps. This involves analyzing the differences in capabilities, processes, and technologies.

5. **Prioritize gaps**: Evaluate the identified gaps based on their impact on the organization's objectives and their urgency. This helps in prioritizing which gaps to address first.

6. **Develop action plans**: For each of the prioritized gaps, develop detailed action plans that outline the steps needed to close the gaps, including resource allocation, timelines, and responsible parties.

How gap assessment results are used

The result of the gap assessment is used to determine strategic plans, resource allocations, implementation roadmaps, and measures of progress and performance for the transformation:

-   **Strategic planning**: The results from gap assessments feed into the strategic planning process, helping to shape the digital transformation strategy by focusing on closing the most critical gaps.

-   **Resource allocation**: Insights from the assessment guide the allocation of resources (budget, personnel, time) to initiatives that address the most significant gaps and offer the highest return on investment.

-   **Implementation roadmap development**: The findings help in creating a detailed roadmap for the digital transformation, outlining the sequence of initiatives to be undertaken to close the identified gaps.

-   **Performance monitoring**: The initial gap assessment provides a baseline against which the progress of the digital transformation can be measured. Subsequent assessments can be conducted to monitor progress and adjust strategies as needed.

Gap assessments are a foundational element of successful digital transformations, offering the insights needed to tailor strategies, prioritize actions, manage risks, and ultimately steer the organization toward its desired future state in a structured and informed manner.

Known good practices for continuous testing

Here are examples of known good practices corresponding to the *Pillars of Practice for Continuous Testing* in *Chapter 1*. These practices can be used to benchmark the current state of an organization's continuous testing practices:

- **Pillar 1: Test automation**:

  - Unit tests are automated for every piece of code.

  - Integration tests are developed to ensure component interactions work as intended.

  - System tests are automated to verify the application behaves as expected.

  - User acceptance tests are automated to confirm the system meets user requirements.

  - Test suites are continuously updated and refactored for relevance and efficiency.

  - API testing is included in the automation strategy for interface reliability.

  - Performance tests are automated and regularly executed to monitor system responsiveness.

- **Pillar 2: Integration with development**:

  - **Test-driven development (TDD)** is practiced, with tests written before code.

  - **Behavior-driven development (BDD)** is employed to ensure alignment with expected behaviors.

  - Automated testing is integrated into each stage of the CI/CD pipeline for immediate feedback.

  - Pair programming or Copilot-enabled pair programming is encouraged to enhance code quality and facilitate instant testing.

  - Feature toggles are used for managing and testing new features.

  - Security testing is integrated early in the development cycle.

  - Static code analysis tools are utilized during development to catch potential issues early.

- **Pillar 3: Test feedback**:

  - Dashboards are set up for immediate visibility of test results.

  - Alerts for test failures are configured for quick response.

  - Trends in test results are analyzed to identify areas for improvement.

  - A feedback loop for test strategy enhancement is established.

  - Test logs and details are made easily accessible for swift issue diagnostics.

  - Visual testing tools are employed for detecting UI changes.

  - AI and machine learning are leveraged for predictive analysis of test outcomes.

- **Pillar 4: Testing metrics**:
  - Defect density is measured to track the number of issues per code unit.
  - Test coverage is calculated to ensure comprehensive testing.
  - **Mean time to resolution (MTTR)** is monitored to assess issue resolution efficiency.
  - Test pass/fail rates are tracked to gauge overall quality.
  - The frequency of test executions is analyzed for test suite optimization.
  - Test efficiency is evaluated to assess the productivity of testing efforts.
  - The defect escape rate is recorded to measure the incidence of issues reaching production.

- **Pillar 5: Risk-based testing**:
  - Testing efforts are prioritized based on critical business functionalities.
  - Security vulnerability priorities are identified using threat modeling and high-priority vulnerability tests are assigned high priority.
  - High-traffic features undergo focused performance testing.
  - Testing resources are allocated based on the complexity and risk of components.
  - Critical system paths are subjected to stress testing.
  - High-risk user scenarios are tested based on user behavior analysis.
  - Areas with a history of frequent issues are prioritized for testing.

- **Pillar 6: Test environment and test data management**:
  - Test environments are automatically set up and torn down.
  - Production-like conditions are replicated in test environments using containerization.
  - Test data privacy is ensured through data masking and anonymization.
  - Version control of test environments is maintained for consistency.
  - Service virtualization is used to simulate unavailable system components.
  - Test data is diverse and comprehensive for accurate testing.
  - Test environments are regularly updated to match production settings.

- **Pillar 7: Collaboration and communication**:
  - Quality assurance alignment is fostered through regular cross-functional team meetings.
  - Transparent testing progress is supported by shared tools for tracking.

- Real-time communication among team members is maintained.

- A blame-free culture is promoted for open discussion of failures and learning.

- Successful testing strategies and learnings are shared across teams.

- Knowledge gaps between developers and testers are bridged through pair-testing.

- Testing knowledge and insights are documented and shared on collaborative platforms.

- **Pillar 8: Continuous learning and adaptation**:

  - Testing tools and processes are regularly reviewed and improved.

  - Latest testing trends are followed through participation in workshops and conferences.

  - Experimentation with new testing methodologies is encouraged.

  - Testing strategies are adapted based on feedback from production issues.

  - Learning from each release cycle is facilitated through retrospectives.

  - The testing team's development is supported by certifications and training.

  - Testing processes are refined continuously using analytics for improvement.

Known good practices for continuous quality

Here are examples of known good practices corresponding to the *Pillars of Practice for Continuous Quality* in *Chapter 1*. These practices can be used to benchmark the current state of an organization's continuous quality practices:

- **Pillar 1: User-centric focus**:

  - User feedback is regularly collected and analyzed to inform quality improvements.

  - Usability testing is conducted consistently to ensure user interface intuitiveness and efficiency.

  - Customer satisfaction surveys are routinely deployed, and results are used to guide development priorities.

  - User experience metrics are monitored continuously to assess and enhance satisfaction.

  - Feedback loops with users are established to quickly identify and address usability issues.

- **Pillar 2: Integrated quality metrics**:

  - Defect rates are tracked across all stages of the software development life cycle.

  - System uptime metrics are monitored to ensure reliability and availability.

  - Performance benchmarks are set and measured to maintain optimal application speed.

- Quality metrics are integrated into the CI/CD pipeline for real-time visibility.
- Code quality assessments are performed regularly to uphold standards.

- **Pillar 3: Proactive quality assurance**:

  - Static code analysis is performed early and often to detect potential issues.
  - Design reviews are conducted at the outset of development cycles to ensure architectural soundness.
  - Architectural evaluations are regularly performed to prevent scalability and performance issues.
  - Code reviews are mandatory for all new commits to maintain high-quality standards.
  - Security vulnerabilities are proactively identified and remediated through automated scanning.

- **Pillar 4: Continuous improvement**:

  - Regular retrospectives are held to reflect on processes and identify improvement opportunities.
  - Continuous learning is encouraged among team members to adopt best practices.
  - Process and tool effectiveness are reviewed continuously for optimization.
  - Quality improvement goals are set and tracked as part of the development process.
  - Innovation in quality practices is fostered to stay ahead of emerging challenges.

- **Pillar 5: Collaboration and communication**:

  - Cross-functional teams collaborate closely to ensure a unified quality approach.
  - Quality goals and metrics are communicated transparently across all departments.
  - Regular sync-ups are held to discuss quality issues and solutions.
  - A common platform is used for tracking and managing quality-related activities.
  - Stakeholders are involved early in the development process to align on quality expectations.

- **Pillar 6: Stable and reliable releases**:

  - Comprehensive testing is conducted prior to each release to ensure stability.
  - Release processes are standardized and automated to minimize human error.
  - Deployment practices include canary releases and blue-green deployments to ensure reliability.
  - Post-release monitoring is implemented to quickly detect and address any issues.
  - Rollback procedures are established and tested for emergency use.

- **Pillar 7: Risk management**:

  - Risk assessments are conducted at regular intervals throughout the project life cycle.

  - Efforts are prioritized based on the potential impact on user satisfaction and system stability.

  - A risk register is maintained and updated to track and manage identified risks.

  - Contingency plans are developed for high-risk scenarios to mitigate potential impacts.

  - Stakeholders are regularly informed about risk status and mitigation strategies.

- **Pillar 8: QA automation**:

  - Automated testing tools are utilized to increase test coverage efficiently.

  - **Continuous integration (CI)** processes include automated quality checks.

  - Automated performance testing is integrated into the development cycle.

  - Regression tests are automated to ensure new changes do not break existing functionality.

  - Automation scripts are regularly reviewed and updated to adapt to new testing needs.

These practices affirm the commitment to integrating quality throughout the development, delivery, and production processes, focusing on enhancing user satisfaction and achieving stable, reliable releases with fewer user issues.

Known good practices for continuous security

Here are examples of known good practices corresponding to the *Pillars of Practice for Continuous Security* in *Chapter 1*. These practices can be used to benchmark the current state of an organization's continuous security practices:

- **Pillar 1: DevSecOps culture**:

  - Security responsibilities are shared among all team members, not isolated to a single group.

  - DevOps practices are integrated with security principles from the outset.

  - Cross-functional training on security practices is provided to all team members.

  - Regular communication on security matters is encouraged across development, operations, and security teams.

  - Security objectives are aligned with overall project goals and metrics.

- **Pillar 2: Security awareness and training**:

  - Security training sessions are conducted regularly for all team members.

  - Awareness programs are developed to keep security top of mind for everyone involved.

  - Security best practices are integrated into onboarding processes for new hires.

  - Continuous learning opportunities in security are provided to keep skills updated.

  - Phishing simulations and other security exercises are conducted to assess team readiness.

- **Pillar 3: Security integration in the lifecycle**:

  - Security requirements are defined at the start of each project.

  - Security tools and practices are embedded into the CI/CD pipeline.

  - Code reviews include security analysis to detect vulnerabilities early.

  - Security checkpoints are established at each stage of the software lifecycle.

  - Collaboration tools are used to integrate security insights into the development workflow.

- **Pillar 4: Automated security testing**:

  - **Software composition analysis (SCA)** and **static application security testing (SAST)** are performed automatically on all code commits.

  - **Dynamic application security testing (DAST)** is conducted regularly throughout the development process.

  - Automated vulnerability scanning is integrated into the build process.

  - Security dependencies are checked for vulnerabilities using automated tools.

  - Container and orchestration images are scanned automatically for misconfigurations and vulnerabilities.

- **Pillar 5: Proactive risk management**:

  - Security risk assessments are conducted at regular intervals.

  - Threat modeling is performed for new features and significant changes.

  - Potential security vulnerabilities are identified and mitigated before deployment.

  - Security controls are reviewed and updated based on the latest threat intelligence.

  - Stakeholders are informed about potential risks and mitigation strategies.

- **Pillar 6: Rapid incident response**:

  - An incident response plan is established and regularly updated.

  - Incident response drills are conducted to ensure team readiness.

  - Security incidents are rapidly identified through continuous monitoring.

  - Automated tools are used to assist in the quick containment of breaches.

  - Post-incident analysis is conducted to learn from security events and improve response.

- **Pillar 7: Continuous monitoring and compliance**:

  - Continuous monitoring solutions are deployed to detect security threats in real time.

  - Security alerts are prioritized based on severity and potential impact.

  - Compliance with security standards and regulations is continuously verified.

  - Security audits are conducted regularly to assess and improve security posture.

  - Compliance documentation is maintained and updated regularly.

- **Pillar 8: Feedback and continuous improvement**:

  - Feedback mechanisms are in place to gather insights from security events.

  - Lessons learned from incidents are shared across teams to prevent recurrence.

  - Security practices are regularly reviewed and updated based on feedback.

  - Evolving security threats are monitored to adapt security strategies accordingly.

  - Continuous improvement cycles are applied to security processes and tools.

These practices assertively outline the proactive, integrated, and continuous approach to security within the framework of continuous security, emphasizing the reduction of security events and their resolution times.

Known good practices for continuous feedback

Here are examples of known good practices corresponding to *the Pillars of Practice for Continuous Feedback* in *Chapter 1*. These practices can be used to benchmark the current state of an organization's Continuous Feedback practices:

- **Pillar 1: Stakeholder and user engagement**:

  - Regular feedback sessions are held with stakeholders and users to gather insights.

  - Surveys and feedback forms are deployed frequently to capture user experiences.

- A feedback portal is established for stakeholders to submit their input anytime.
- Feedback from social media and forums is actively monitored and collected.
- Stakeholder and user workshops are conducted to dive deeper into feedback.

- **Pillar 2: Feedback integration in the development**:
  - Feedback is directly linked to the development backlog for prioritization.
  - A process for rapid feedback triage and integration into development cycles is established.
  - Development teams hold regular meetings to review and act on feedback.
  - Automated tools are used to streamline feedback integration into the development pipeline.
  - Feedback metrics are tracked to inform development priorities and decisions.

- **Pillar 3: Rapid iteration and implementation**:
  - Development cycles are shortened to facilitate quicker feedback implementation.
  - Continuous deployment practices are adopted to accelerate release cycles.
  - A/B testing is utilized to quickly iterate based on user feedback.
  - Feedback implementation effectiveness is reviewed in sprint retrospectives.
  - Prototyping and MVPs are used to test and refine ideas based on feedback.

- **Pillar 4: Data-driven decision making**:
  - Feedback is quantitatively analyzed to identify trends and priorities.
  - User behavior analytics tools are employed to understand feedback context.
  - Decision-making processes are structured around feedback analysis outcomes.
  - A feedback dashboard is maintained to visualize feedback data for informed decisions.
  - Sentiment analysis is applied to qualitative feedback for deeper insights.

- **Pillar 5: Feedback transparency and communication**:
  - Feedback actions and decisions are communicated back to stakeholders regularly.
  - A public roadmap is maintained to show how feedback influences development.
  - Forums and community channels are used to discuss feedback and related actions.

- Regular updates on feedback implementation progress are provided.
- Challenges and lessons learned from feedback are shared openly with stakeholders.

- **Pillar 6: Continuous learning and adaptation**:

  - Feedback insights are incorporated into continuous improvement processes.
  - Teams are encouraged to experiment based on feedback to find optimal solutions.
  - Things learned from feedback is shared across teams to foster organizational growth.
  - Feedback loops are continuously refined for efficiency and effectiveness.
  - Adaptation of processes and practices is based on ongoing feedback analysis.

- **Pillar 7: Measuring impact and effectiveness**:

  - Metrics related to feedback implementation speed are tracked and optimized.
  - The impact of feedback on system reliability is continuously monitored.
  - User satisfaction surveys are analyzed to measure feedback effectiveness.
  - Changes in release frequency and recovery times due to feedback are evaluated.
  - Feedback loop effectiveness is reviewed through regular audits and assessments.

- **Pillar 8: Balancing speed and quality**:

  - Quality assurance processes are integrated with rapid feedback implementation strategies.
  - Feedback is prioritized based on the potential impact on system quality and reliability.
  - Automated testing is scaled up to maintain quality while implementing feedback quickly.
  - Peer reviews and pair programming are used to balance speed with quality in feedback implementation.
  - Release gates are established to ensure feedback implementations meet quality standards.

These practices affirm a strategic approach to utilizing stakeholder and user feedback for enhancing system reliability, accelerating release frequency, and improving recovery times, always emphasizing the importance of maintaining quality.

Figure 6.5 illustrates an example of a gap assessment tool for continuous testing that can take the output from the survey and calculate a gap score for each of the practices.

Continuous Testing (CT) Practices	(I) Importance	(P) Practice Level	(G) GAP
Unit tests are automated for every new piece of code.	4	3	4
Integration tests are developed to ensure component interactions work as intended.	3	3	3
System tests are automated to verify the application behaves as expected.	4	1	12
User acceptance tests are automated to confirm the system meets user requirements.	4	1	12
Test suites are continuously updated and refactored for relevance and efficiency.	3	1	9
API testing is included in the automation strategy for interface reliability.	4	1	12
Performance tests are automated and regularly executed to monitor system responsiveness.	4	2	8
Test Automation	3.7	1.7	8.6
	Importance Average	Practice Level Average	Average GAP

Figure 6.5 – Gap assessment tool

As data is entered for the importance and practice level values, the tool creates a summary of the gap scores and ranks in a summary format shown in *Figure 6.4*. The same tool can be used to determine gaps for continuous testing, quality, security, and feedback when the practices for each topic are loaded into the tool.

Understanding CSVSM

Current State Value Stream Mapping (**CSVSM**) in the context of assessments, discovery, and benchmarking stages of digital transformations, specifically relating to continuous testing, quality, security, and feedback, is a strategic and analytical tool used to visualize and understand the flow of information, and processes required to deliver a product or service to the customer.

Figure 6.6 is an example of a CSVSM:

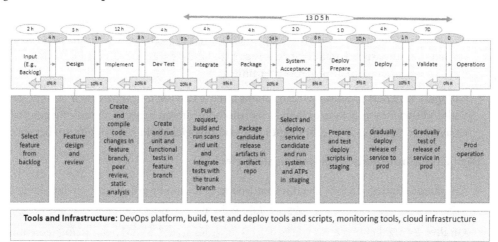

Figure 6.6 – Sample CSVSM

In the realm of digital transformations, particularly focusing on continuous testing, quality, security, and feedback, CSVSM plays a crucial role in identifying inefficiencies, bottlenecks, and areas for improvement in existing processes and systems. Here's how CSVSM is applied in these contexts:

- **Assessment**:

  - **Identify current processes**: CSVSM helps in the detailed mapping of existing testing, quality assurance, security, and feedback collection processes. This step is crucial for understanding how tasks are currently performed, the sequence of these tasks, and the interaction between different roles and technologies.

  - **Pinpoint inefficiencies**: By visually representing the current state, inefficiencies such as redundant processes, delays, and areas where quality or security may be compromised become evident. This visibility is essential for targeting improvements.

- **Discovery**:

  - **Highlight improvement areas**: The discovery phase involves analyzing the CSVSM to identify areas where digital tools and methodologies can be introduced or optimized. This includes integrating automated testing, enhancing security protocols, improving quality checks, and streamlining feedback mechanisms.

  - **Stakeholder engagement**: CSVSM facilitates discussions among stakeholders by providing a clear depiction of current processes and highlighting areas for digital enhancement. This aids in aligning the transformation goals with business objectives and user needs.

- **Benchmarking**:

  - **Establish performance metrics**: CSVSM serves as a baseline to measure the effectiveness of digital transformation initiatives. By establishing **key performance indicators** (**KPIs**) related to testing speed, quality levels, security standards, and feedback loop efficiency, organizations can monitor progress over time.

  - **Compare with best practices**: Organizations can use CSVSM to compare their current state with industry best practices or standards. This comparison helps in identifying gaps and areas where adopting new technologies or methodologies could yield significant benefits.

- **Digital transformations to continuous strategies**:

  - **Continuous testing**: CSVSM identifies bottlenecks in the testing process that can be addressed through automation, enabling faster and more frequent testing cycles.

  - **Quality**: By highlighting areas where defects or issues arise, CSVSM helps in implementing quality improvement measures such as more rigorous testing protocols or quality checks at earlier stages in the development process.

  - **Security**: Security vulnerabilities can be pinpointed through CSVSM, leading to the adoption of enhanced security practices and tools throughout the development life cycle.

  - **Feedback**: CSVSM can reveal delays or inefficiencies in gathering and responding to feedback, guiding the implementation of more effective feedback loops that promote continuous improvement.

In summary, CSVSM is a foundational tool in the digital transformation journey, especially in areas focused on continuous testing, quality, security, and feedback. It provides a comprehensive view of existing processes, facilitating targeted improvements, stakeholder engagement, and the effective benchmarking of progress towards digital maturity.

Steps to creating a CSVSM

Creating a CSVSM for continuous testing involves several steps aimed at understanding and visualizing the existing testing processes, identifying inefficiencies, and highlighting opportunities for implementing continuous testing practices. Here's a concise guide to the steps involved:

1. **Define the scope and objectives**:

   - **Scope**: Determine the boundaries of the process to be mapped, focusing on the testing activities within the software development life cycle.

   - **Objectives**: Clearly state what you aim to achieve with continuous testing, such as reducing time-to-market, improving test coverage, or enhancing quality.

2. **Gather a cross-functional team**:

   • Assemble a team that includes members from different areas involved in the testing process, such as developers, testers, QA managers, and operations staff. This ensures a comprehensive understanding of the process from multiple perspectives.

3. **Map the current process**:

   • **Data collection**: Collect data on the current testing process, including tools used, sequence of activities, time taken for each activity, and handoffs between teams.

   • **Visual representation**: Use a value stream mapping tool or simply pen and paper to create a visual representation of the current testing process. Mark each step in the process, along with metrics such as duration, waiting time, and the frequency of tests.

4. **Identify and mark value-adding and non-value-adding steps**:

   • Analyze each step to determine whether it adds value to the final product from the customer's perspective. Mark non-value-adding steps (waste) such as delays, unnecessary handoffs, or redundant testing.

5. **Identify bottlenecks and areas of delay**:

   • Look for steps in the process where delays occur, where there is a buildup of work, or where bottlenecks restrict the flow of work. These are critical areas that impact the efficiency of the testing process.

6. **Collect feedback**:

   • Engage with the team and other stakeholders to gather insights and feedback on the current testing process. This can provide valuable context and help in identifying areas for improvement not evident in the initial mapping.

7. **Analyze the map for opportunities**:

   • With the current state clearly laid out, identify opportunities for introducing continuous testing practices. This includes automating repetitive tasks, improving test data management, integrating testing earlier in the development process (shift-left), and enhancing collaboration between developers and testers.

8. **Plan for the future state**:

   • Based on the analysis, sketch a future state value stream map that incorporates continuous testing practices aimed at addressing the identified inefficiencies and meeting the objectives set in *Step 1*.

Creating a CSVSM for continuous testing is a critical step in transitioning to a more efficient, effective, and continuous testing process. It helps organizations visualize their current practices, identify inefficiencies, and plan targeted improvements to enhance quality and reduce time-to-market.

Challenges to overcome with value stream mapping

CSVSM is a vital tool in understanding and optimizing processes by visualizing the flow of information, materials, and work through a system. However, implementing CSVSM, especially in complex environments such as digital transformations toward continuous testing, quality, security, and feedback, comes with its set of challenges and hurdles. Recognizing these challenges is the first step toward devising effective strategies to overcome them.

Challenges and hurdles

Let's understand the challenges with CSVSM:

- **Complexity of processes**: Many organizations have processes that are highly complex, intertwined, and sometimes undocumented, making it difficult to accurately map the current state.

- **Stakeholder engagement**: Securing engagement and cooperation from all relevant stakeholders can be challenging, especially in large or siloed organizations where different departments may have competing priorities.

- **Data overload**: The sheer volume of data related to processes, delays, and inefficiencies can be overwhelming, making it hard to identify the most critical issues that need addressing.

- **Resistance to change**: Employees and management may resist changes suggested by the value stream mapping exercise, especially if they are comfortable with the status quo or fearful of the implications of change.

- **Lack of expertise**: Properly conducting a CSVSM exercise requires a certain level of expertise and experience, which may not be present internally in the organization.

Suggestions to overcome the hurdles

Here are a few ways to overcome the challenges with CSVSM:

- **Simplify and prioritize**: Break down complex processes into smaller, manageable sections and prioritize them based on their impact on the overall value stream. This makes mapping more manageable and focused.

- **Engage and communicate**: Foster a culture of transparency and inclusivity by actively engaging stakeholders from the beginning. Clearly communicate the benefits of CSVSM and how it can positively impact their work.

- **Leverage technology**: Use software tools and platforms that can help manage and analyze the data effectively. Generative AI, for instance, can automate part of the data collection and analysis process, highlighting key areas of focus.

- **Address resistance proactively**: Implement change management strategies that address fears and resistance head-on. Include training, workshops, and clear communication about the changes and their benefits to both the organization and its employees.

- **Seek external expertise**: If internal expertise is lacking, consider hiring external consultants who specialize in CSVSM and can guide the organization through the process. This can also help in transferring knowledge and skills to internal teams.

- **Iterative approach**: Treat CSVSM as an iterative process rather than a one-time project. Continuous improvement should be the goal, allowing for adjustments and refinements as more information is gathered and the digital transformation progresses.

- **Celebrate successes**: Acknowledge and celebrate improvements and successes achieved through CSVSM. This boosts morale and demonstrates the tangible benefits of embracing change.

By acknowledging these challenges and strategically addressing them, organizations can effectively utilize CSVSM to facilitate digital transformations, enhancing continuous testing, quality, security, and feedback mechanisms. This approach not only streamlines processes but also fosters a culture of continuous improvement and innovation.

How generative AI can be used to accelerate discovery and benchmarking

Generative AI can significantly enhance the efficiency and quality of current state discovery, benchmarking gap assessments, and the creation of CSVSM in several impactful ways:

- **Automated data collection and analysis**:

  - **Efficiency**: Generative AI can automate the collection of data across various systems and tools used in the current state processes. By parsing through logs, project management tools, and development environments, AI can quickly gather necessary information, reducing manual effort and time.

  - **Quality**: AI algorithms can analyze this data more accurately and consistently than manual methods, identifying patterns, bottlenecks, and inefficiencies that might not be evident to human analysts.

- **Enhanced visualization of value streams**:

  - **Efficiency**: AI can automatically generate visual representations of value streams based on the collected data. This automation speeds up the creation of CSVSMs, allowing teams to focus on analysis and improvement strategies rather than on the manual drawing of maps.

  - **Quality**: Generative AI can produce more detailed and dynamic visualizations, incorporating real-time data and analytics directly into the value stream map. This provides a deeper understanding of process flows, variations, and inefficiencies.

- **Predictive analytics for benchmarking and gap assessments**:

  - **Efficiency**: Through machine learning models, AI can predict outcomes based on current processes and compare them with industry benchmarks or desired performance levels. This accelerates the benchmarking and gap assessment phases by providing immediate insights into where gaps exist and their potential impact.

  - **Quality**: AI's predictive capabilities allow for more accurate forecasting of improvements and their expected benefits. By simulating changes in the value stream map, organizations can prioritize interventions that offer the highest return on investment.

- **Natural language processing (NLP) for enhanced discovery**:

  - **Efficiency**: NLP can analyze qualitative data from interviews, open-ended survey responses, and documentation to extract insights about the current state that might not be captured through quantitative methods alone.

  - **Quality**: This deep analysis can uncover underlying issues, stakeholder perceptions, and hidden opportunities for improvement that are crucial for a comprehensive understanding of the current state and for crafting targeted improvement strategies.

- **Continuous improvement feedback loops**:

  - **Efficiency**: Generative AI can facilitate continuous monitoring of the process changes implemented based on the CSVSM analysis. By continuously analyzing performance data, AI can suggest real-time adjustments to processes, ensuring that the value stream remains optimized.

  - **Quality**: This ongoing optimization helps maintain high-quality standards and adapt quickly to changes in the business environment or technology landscape.

- **Collaboration and stakeholder engagement**:

  - **Efficiency**: AI-driven platforms can enhance collaboration among team members by providing a centralized, interactive environment for analyzing and discussing the value stream map. This improves the efficiency of stakeholder meetings and decision-making processes.

- **Quality**: Improved collaboration ensures that diverse perspectives are considered in the analysis, leading to more comprehensive and effective improvement strategies.

In summary, generative AI offers powerful tools for automating and enhancing the processes of current state discovery, benchmarking, gap assessments, and the creation of CSVSMs. By leveraging AI's capabilities, organizations can achieve a more accurate, detailed, and efficient analysis of their current state, leading to targeted and effective improvements in their value streams.

Summary

In the transformative journey towards achieving excellence in continuous testing, quality, security, and feedback within digital environments, the pivotal role of comprehensive assessments cannot be overstated. This chapter delves into the nuanced methodologies of interviews, surveys, gap assessments, and value stream mapping as foundational tools for discovery and benchmarking. Through engaging interviews and meticulously designed surveys, organizations can capture the nuanced insights of stakeholders, ranging from frontline developers to top-level management. These tools not only illuminate the current state of practices and processes but also foster a culture of inclusivity and open communication, paving the way for targeted improvements.

Gap assessments stand out as a critical analytical tool, providing a clear demarcation between the current state and the desired future state of digital practices. By systematically identifying areas of improvement, organizations can prioritize initiatives that yield the highest impact on quality, security, and feedback mechanisms. Complementing this, value stream mapping offers a visual representation of the flow of information and processes, highlighting inefficiencies and bottlenecks that impede progress. This comprehensive approach ensures that no stone is left unturned in the quest for digital excellence.

The advent of generative AI marks a significant leap forward in enhancing the efficiency and quality of these assessments. By automating data collection, analysis, and even the generation of value stream maps, AI technologies free up valuable human resources to focus on strategic decision-making and implementation. Moreover, AI's predictive analytics capabilities offer unprecedented insights into potential future states, enabling organizations to make data-driven decisions with greater confidence. This synergy between human expertise and AI's computational power accelerates the pace of digital transformation, ensuring that organizations remain competitive in an ever-evolving landscape.

As we move forward, the next chapter promises to delve deeper into the technological backbone that supports continuous testing, quality, security, and feedback. We will explore a variety of technology frameworks and tools that are reshaping the digital domain.

7
Selecting Tool Platforms and Tools

In the rapidly evolving landscape of software development, the emphasis on continuous improvement across testing, quality, security, and feedback has never been more critical. This chapter serves as a comprehensive guide to understanding tool platforms and tools that enable developers and organizations to maintain excellence through automation and efficient practices. The chapter demystifies what tool platforms and tools are, delving into their core definitions, functionalities, and the pivotal role they play in streamlining software workflows.

This chapter provides you with a deep understanding of how each platform and tool can be leveraged to foster a culture of continuous improvement and resilience in the face of ever-changing technological challenges.

Selecting the right tool platforms and tools is a critical decision that can significantly impact the efficiency and effectiveness of development processes and pipelines. This chapter describes a methodology and tools for making informed decisions when choosing tool platforms and tools. This methodology is designed to guide you through evaluating your specific needs, the strengths, and weaknesses of different options, and aligning your choices with your strategic goals.

By the end of this chapter, you will have gained a comprehensive understanding of the fundamental concepts of tool platforms and tools in software development. We'll also discuss the diverse array of tool platforms and tools available for continuous testing, quality assurance, security, and feedback.

Finally, we'll learn about the strategic approach to selecting the most suitable platforms and tools for your projects, ensuring they align with your objectives and enhance your development life cycle.

This chapter is organized under the following headings:

- Tool platforms and tools concepts
- Platforms and tools for continuous testing, quality, security, and feedback
- Source of platforms and tools

- Factors for comparing tool platforms and tools
- Methodology for selecting tool platforms and tools

Let's get started!

Tool platforms and tools concepts

Tool platforms and tools are closely related yet distinct concepts within the software development ecosystem, each playing a crucial role in streamlining and enhancing various processes through their orchestration, automation, and integration capabilities. Here's how they relate to each other.

Tool platforms

These are comprehensive, structured environments that provide a foundation for developing, testing, deploying, and managing software applications. Platforms define a standardized way to build and deploy applications by offering a set of guidelines, coding standards, libraries, and helper tools. They aim to simplify the development process by reducing the need for repetitive actions, promoting code reuse, and providing a set of common practices and conventions. A legitimate tool platform should support the ability to integrate with other tools. Platforms encompass not only development tools but also runtime environments, services, and APIs that applications and other tools can use. A platform can host other tools and provide data services, APIs, analytics, and more, facilitating the running, and management of tools applications in a unified environment.

Let's look at the characteristics:

- **Extensibility**: Platforms are designed to be extended and integrated with a wide range of applications and services. They provide a set of tools and services that applications can leverage.
- **Ecosystem**: This often includes a marketplace or a community where third-party services and integrations can be added to enhance functionality.
- **Flexibility**: While they provide many tools and services, platforms generally offer more flexibility in how applications are developed, deployed, and managed.

Tools

In the context of software development, tools are specific software or applications that perform distinct tasks or functions. These can range from compilers, debuggers, and code editors to more specialized tools for version control, project management, continuous integration, and testing. Tools are often designed to be integrated into larger development workflows, and they can be standalone, part of a broader ecosystem, or integrated with tool platforms.

Relationship between tool platforms and tools

Figure 7.1 illustrates the relationships between tool platforms, tools, and target applications.

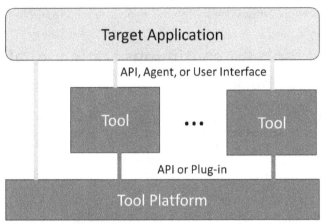

Figure 7.1 – Relationships between tool platforms, tools, and applications

The relationships between tool platforms, tools, and applications can be explained with a look at capabilities for foundation integration, standardization and efficiency, how they complement each other and how they enhance each other, as described in the following paragraphs.

- **Foundation and integration**: Tool platforms serve as the foundation or the ecosystem within which tools operate. They often come with built-in tools or are designed in a way that allows the seamless integration of external tools. For instance, a web development platform might include built-in tools for templating and database migration but also allow the integration of external testing tools or development environments.

- **Standardization and efficiency**: Platforms provide the guidelines and standards that tools adhere to, ensuring consistency and efficiency in development practices. Tools, when used within these platforms, help automate repetitive tasks, enhance productivity, and ensure that development practices meet the standards set by the Platforms.

- **Complementarity**: Tools complement platforms by providing specific functionalities that the platforms might not cover out of the box. For example, while the platforms might dictate how applications should be structured and provide basic components for building applications, tools such as version control systems, continuous integration pipelines, and testing suites complement the platforms by adding functionality for team collaboration, code integration, and quality assurance.

- **Enhancement**: Tools can enhance the capabilities of platforms by adding new features, improving performance, or making development workflows more flexible and efficient. Developers often choose tools that are known to work well with their chosen platforms to ensure a smooth development process.

Tool platforms and tools are interdependent, with platforms providing the structured environment and standards for development, and tools offering the specific functionalities needed to carry out development tasks efficiently within those platforms. This relationship allows for the streamlined development of software applications that are robust, maintainable, and scalable.

Platforms and tools for continuous testing, quality, security, and feedback

Platforms and tools for continuous testing, quality, security, and feedback, although distinct in their primary focus, are interrelated and often overlap in practice. The following sections explain how they can be distinguished and how they complement each other.

Continuous testing platforms and tools

Continuous testing is distinguished by its focus on automated testing to validate that software meets the pass criteria for each test case, as it is developed and delivered. It's iterative and integrated into the development pipeline.

- **Tools and tool platforms**: This includes automated testing tools that cover all types of testing including unit testing, integration testing, system testing, and performance testing.

- **Complements**: It ensures that new features and changes do not introduce regressions, by identifying defects early. Continuous feedback is incorporated by integrating user and stakeholder feedback into tests to ensure that what is developed aligns with expectations.

Many tools can be classified as tool platforms and tools for continuous testing. Some examples are given here:

- **Tool platforms**:

  - **Jenkins**: While it's often categorized as a **continuous integration and continuous deployment (CI/CD)** tool, Jenkins can be seen as a platform for continuous testing due to its extensive plugin ecosystem that integrates with a wide range of testing tools, allowing for automated testing as part of the build and deployment process.

  - **TestRail**: This offers comprehensive test case management to organize, manage, and track the testing process. It integrates with many issue-tracking and automation tools, enhancing continuous testing workflows.

- **Tools**:

  - **Selenium**: An automated testing tool for web applications that can be integrated into various development environments and platforms for end-to-end automation.

- **Postman**: While primarily a tool for API testing, Postman can be integrated into tool platforms for automated API tests, contributing to continuous testing efforts.

Continuous quality platforms and tools

Continuous quality is distinguished by its emphasis on integrating quality metrics throughout all processes. It's not only about finding defects but also about preventing them by incorporating quality standards from the start.

- **Tools and platforms**: These include quality management systems, code quality metrics tools such as SonarQube, and monitoring tools that track quality metrics over time.

- **Complements**: It benefits from continuous testing tools that provide immediate feedback on the quality of the code. Continuous feedback mechanisms are key to understanding user expectations for quality. Continuous security is a subset of quality, so the tools that ensure security also contribute to overall quality.

Many tools can be classified as tool platforms and tools for continuous quality. Some examples are given here:

- **Tool platforms**:

  - **SonarQube**: It stands as a quality framework with its ability to integrate with development environments, CI/CD pipelines, and version control systems to continuously inspect code quality and security vulnerabilities.

  - **Codacy**: A platform that automatically identifies issues through static code analysis. It integrates with GitHub, GitLab, and Bitbucket, supporting a collaborative and integrated approach to code quality.

- **Tools**:

  - **Coverity**: A static analysis tool that identifies software vulnerabilities and quality issues that can be integrated into CI/CD pipelines for continuous analysis.

  - **ESLint**: As a tool, it's focused on linting and fixing problems in JavaScript code, but it can be integrated into various development workflows for continuous code quality checks.

Continuous security platforms and tools

Continuous security is distinguished by its focus on integrating security aspects into all stages of development. It's proactive, aiming to prevent security issues before they occur.

- **Tools and platforms**: These include **static application security testing (SAST)**, **dynamic application security testing (DAST)**, and tools for managing credentials and secrets, such as HashiCorp Vault.

- **Complements**: Security is a part of quality; hence, continuous security tools are essential for continuous quality. They also benefit from continuous testing tools for automated security tests. Continuous feedback can provide insights into security concerns from users and stakeholders.

Many tools can be classified as tool platforms and tools for continuous security. Some examples are given here:

- **Tool platforms**:

  - **OWASP Zed Attack Proxy (ZAP)**: It can be integrated into CI/CD pipelines for automated security testing, making it a part of a broader security framework.

  - **Metasploit Framework**: Beyond being a tool for exploitation, it integrates with various penetration testing tools and environments, facilitating a comprehensive security testing and validation framework.

- **Tools**:

  - **Nmap**: While Nmap is a tool for network exploration and security auditing, it's often integrated into security frameworks as part of automated network scanning and monitoring.

  - **Burp Suite**: Primarily a tool for web application security testing, it offers extensibility through its API and integrations, allowing it to be part of a larger security assessment framework.

Continuous feedback platforms and tools

Continuous feedback is distinguished by its involvement in collecting and implementing stakeholder and user feedback into the development cycle to improve the product continuously.

- **Tools and platforms**: These include user feedback platforms, feature flagging tools such as **LaunchDarkly**, and monitoring tools that can provide insights into how users are interacting with the application.

- **Complements**: Feedback can drive improvements in continuous testing by highlighting areas that need additional coverage. It directly informs continuous quality initiatives by highlighting what users care about most. For continuous security, user feedback might uncover perceived or actual security concerns, which can then be addressed.

Many tools can be classified as tool platforms and tools for continuous feedback. Some examples are given here:

- **Tool platforms**:

  - **Salesforce Service Cloud**: Beyond being a popular **Customer Relationship Management (CRM)** tool, it offers a platform that integrates various feedback and support tools, enabling comprehensive management of customer feedback across channels.

  - **Zendesk**: This provides a customer service and engagement platform that integrates with multiple feedback, support, and analytics tools, creating a cohesive system for gathering and acting on customer insights.

- **Tools**:

  - **Typeform**: A tool for creating interactive surveys and forms, which can be integrated into websites or platforms for direct feedback collection.

  - **Hotjar**: Offers tools for feedback through heatmaps, session recordings, and surveys. While primarily a tool, it supports integrations that enable it to fit into broader feedback and analytics frameworks.

Overlap and integration

Many tools serve multiple purposes. For example, a comprehensive continuous testing tool might include security testing features, thereby also serving continuous security needs.

Similarly, project management tools that facilitate continuous feedback might have integrated quality and testing dashboards.

DevOps platforms such as GitLab or Jenkins can orchestrate a pipeline that includes stages for continuous testing, quality, security, and collecting feedback.

In summary, while tools and platforms can be categorized based on their primary function, the most effective development and delivery pipelines will integrate these categories, recognizing that they are not silos but parts of a cohesive whole. The goal is to develop software that is not only functional but also of high quality, secure, and well-aligned with user needs and expectations.

Source of platforms and tools

One of the most critical decisions to make when selecting platforms and tools is to decide the source of each platform and tool. *Figure 7.2* indicates the advantages and disadvantages of open-source tools versus vendor product tools versus **do-it-yourself** (**DIY**) tools for continuous testing, quality, security, and feedback.

Open-Source Tool Platforms & Tools	Vendor Product Tool Platforms & Tools	DIY (Home Grown) Tool Platforms & Tools
Advantages: • Cost-effective • Community support • Flexibility & customization • Transparency	**Advantages:** • Professional support • Reliability • Ease-of-use • Integration	**Advantages:** • Fully customized • Control • Integration
Disadvantages: • Support and maintenance • Learning curve • Integration • Reliability	**Disadvantages:** • Cost • Lock-in • Less customization • Opague	**Disadvantages:** • Resource Intensive • Support • Sustainability • Risk

Figure 7.2 – Source of tool platforms and tools

Let's explore them in detail.

Open-source tools

In the following paragraphs, we will explore open-source tools for continuous testing, quality, security, and feedback.

Open-source continuous testing tools

Open-source continuous testing tools allow teams to integrate testing seamlessly into the software development and deployment process without the cost constraints of proprietary software. These tools enable automated testing at every stage of development, ensuring errors are detected early and addressed promptly to maintain software quality.

Example: Selenium is a widely used open-source tool for automating web browsers, enabling developers to write test cases that simulate user interactions across various browsers and platforms.

Open-source quality tools

Open-source quality tools provide a cost-effective solution for maintaining high standards in software development. They allow for static code analysis, monitoring of code complexity, and adherence to coding standards, helping maintain clean and efficient code bases.

Example: SonarQube is an open-source platform that continuously inspects code quality, using static analysis to automatically review code and detect bugs, code smells, and security vulnerabilities.

Open-source security tools

Open-source security tools provide essential security capabilities such as penetration testing and vulnerability scanning at no cost, enabling organizations to strengthen their application security posture effectively. These tools are often community-driven, benefiting from contributions that keep them up to date with the latest security threats.

Example: OWASP ZAP is a leading open-source security tool developed by one of the most reputable names in web security, OWASP, designed to find security vulnerabilities in web applications during the development and testing phases.

Open-source feedback tools

Open-source feedback tools help organizations gather and manage user or stakeholder feedback efficiently, facilitating continuous improvements in product development. These tools integrate into development workflows, allowing teams to collect and respond to user insights and expectations directly.

Example: MantisBT is an open-source issue tracker that offers a simple yet powerful platform for tracking user feedback and issues, helping organizations refine and enhance their products based on real user experiences.

Advantages of open-source tools

- **Cost effectiveness**: Open-source tools are usually free to use, which can significantly reduce costs.
- **Community support**: There's often a large community for support and development, which can help in troubleshooting and improving the tools.
- **Flexibility and customization**: Open-source allows for customization to fit specific needs.
- **Transparency**: The open nature means you can inspect and modify the code, which can be beneficial for security audits.

Disadvantages of open-source tools

- **Support and maintenance**: Lack of formal support can be a challenge; you often rely on community support, which can be variable.
- **Learning curve**: They can have a steep learning curve, especially for teams not familiar with the technology.
- **Integration**: It may require more effort to integrate with existing systems, especially if they are proprietary.
- **Reliability**: Some open-source projects may not be as reliably maintained as commercial products.

Vendor product tools

Vendor product tools for continuous testing provide advanced integrated solutions that support automated testing throughout the software development and deployment process. These commercial tools often offer enhanced support, scalability, and additional features tailored to enterprise needs, ensuring thorough testing and quality assurance.

Example: TestComplete is a commercial automated testing platform that enables teams to create, manage, and run tests across desktop, web, and mobile applications. TestComplete offers a rich set of features including GUI testing, script or scriptless test creation, and integration with other tools in the DevOps pipeline.

Quality product tools

Commercial quality tools provide robust solutions for maintaining high standards in software code quality. These tools typically offer comprehensive analytics, extensive integrations with development environments, and professional support, helping teams adhere to best practices and improve code health.

Example: Coverity is a static code analysis tool that helps developers write safer and more secure code by detecting and resolving defects and vulnerabilities in the code base before they become issues in production.

Security product tools

Vendor product security tools offer sophisticated security solutions that include advanced features such as automated threat detection, real-time protection, and incident response capabilities. These tools are backed by professional support and continuous updates to tackle evolving security threats.

Example: Veracode is a scalable, policy-based application security platform that provides automated cloud-based services for securing web, mobile, and third-party enterprise applications throughout the software development lifecycle.

Feedback product tools

Commercial feedback tools are designed to efficiently capture and manage user feedback across multiple channels. These tools often provide comprehensive analytics, integration with CRM systems, and tools to engage directly with users, enhancing the ability to act on user insights to improve product development.

Example: UserVoice is a product feedback management software that helps businesses gather and analyze customer feedback and prioritize feature requests to better align product development with user needs.

Advantages of vendor product tools

- **Professional support**: These tools often come with professional support and training.

- **Reliability**: Vendors have a commercial interest in maintaining and updating their products.

- **Ease of use**: They are typically designed with user-friendliness in mind, often providing a smoother onboarding process.

- **Integration**: They may offer better integration with other tools and services, particularly within the same ecosystem.

Disadvantages of vendor product tools

- **Cost**: These can be expensive, especially with ongoing licensing fees.

- **Lock-in**: There's the potential for vendor lock-in, making it hard to switch to other tools or vendors.

- **Less customization**: They may not be as customizable as open-source options.

- **Opaque**: The closed-source nature means you cannot inspect or modify the code, which might be a concern for security-sensitive environments.

DIY or home-grown tools

Do-It-Yourself, or home-grown tools can be an attractive choice if an organization does not have budget to buy a vendor product tool and does not want to use open-source tools.

DIY continuous testing tools

DIY continuous testing tools are custom-developed solutions tailored to specific project or organizational needs. These home-grown tools allow teams to integrate unique testing processes and automation frameworks directly into their development environments, offering full control over testing features and integration capabilities.

Example: Custom Automation Framework is a bespoke testing framework built using open-source libraries such as Selenium or Appium, modified and extended with custom scripts and plugins to meet the specific testing requirements of an organization's software projects.

DIY quality tools

DIY quality tools are developed in-house to maintain code standards and quality specific to the development practices of an organization. These tools can be specifically designed to integrate with existing workflows, providing targeted insights and analytics on code quality.

Example: Internal Code Review Tool is an in-house developed tool that integrates with version control systems to automate code reviews and quality checks, tailored to enforce the organization's coding standards and practices.

DIY security tools

Home-grown security tools are built to address specific security needs and integrate seamlessly with an organization's existing IT infrastructure. These tools can be customized to provide specialized security monitoring, threat detection, and incident response tailored to the unique environment and security policies of the company.

Example: Custom Security Dashboard is a self-developed comprehensive dashboard that aggregates security alerts, logs, and data from various sources within the organization, equipped with custom algorithms to detect anomalies and potential threats based on historical data.

DIY feedback tools

DIY feedback tools allow organizations to create customized systems for collecting and managing user feedback directly within their product ecosystems. These tools can be tailored to capture specific types of user input and integrate deeply with product features to provide actionable insights.

Example: Internal Feedback Portal is a custom-built internal portal designed to collect user feedback across various stages of product interaction, integrated with the organization's product management and issue-tracking systems to facilitate direct user engagement and faster response times.

Advantages of DIY tools

- **Fully customized**: This can be tailored exactly to the specific needs of an organization.
- **Control**: Full control over the development, deployment, and maintenance processes.
- **Integration**: This can be designed to integrate seamlessly with the existing infrastructure.

Disadvantages of DIY tools

- **Resource intensive**: Developing and maintaining custom tools requires significant time and expertise.
- **Support**: No external support; the organization is responsible for troubleshooting any issues.
- **Sustainability**: May not be sustainable in the long term as technology evolves and internal resources change.
- **Risk**: High risk if the internal team lacks the necessary expertise.

Each type of tool has its place depending on the needs, resources, and goals of an organization. It's common for businesses to use a mix of these tools to balance the benefits and drawbacks of each. For continuous testing, quality, security, and feedback, the right choice will depend on factors such as the importance of customization, the available budget, the need for support, and the desired level of control over the tools and processes.

Factors for comparing tool platforms and tools

The following are factors that are important to consider when selecting tool platforms and tools for continuous testing, quality, security, and feedback:

- **Compatibility and integration**: Tools should easily integrate with your existing workflow, other tools in use, and the overall IT infrastructure. Seamless integration helps in automating the pipeline and reduces friction.

- **Scope and coverage**: Determine the scope of testing, quality, security, and feedback that the tool or platform provides. The broader the coverage, the more likely it will meet various needs as your projects evolve.

- **Usability and learning curve**: Consider how easy the tool is to use and the time required to train staff. Tools that are user friendly and easy to learn can be adopted more quickly.

- **Automation capabilities**: For continuous processes, automation is key. Evaluate the extent to which a tool can automate repetitive tasks and integrate into CI/CD pipelines.

- **Reporting and analytics**: The ability to generate insightful reports and analytics is crucial for continuous improvement. Tools should offer comprehensive reporting features for tracking progress and making data-driven decisions.

- **Community and support**: Consider the community around open-source tools or the support offered by vendors. A strong community or good support can be invaluable for problem-solving and best practices.

- **Costs**: Assess the total cost of ownership, not just the initial price. This includes licensing fees, maintenance, support, and training costs for vendor products, and the development and maintenance costs for homegrown solutions.

- **Performance and scalability**: The chosen tools should perform well under the expected load and be scalable to grow with your needs without a significant increase in cost or complexity.

- **Security and compliance**: Ensure the tool or platform does not introduce security vulnerabilities and that it complies with relevant industry standards and regulations.

- **Flexibility and customizability**: The ability to customize the tool to suit specific project requirements can be a significant advantage, especially for complex projects.

- **Vendor stability and roadmap**: When selecting vendor tools, consider the vendor's stability, reputation, and roadmap for the product to ensure it will continue to meet your needs in the future.

- **Update frequency and maintenance**: Understand how often the tool is updated. Frequent updates can indicate active development but also consider the effort required to keep up with these updates.

- **Feedback mechanisms**: For tools focused on feedback, assess how they gather, analyze, and manage feedback. The tool should facilitate actionable insights to improve the development process.

- **Cloud native**: This refers to whether the tool is designed for and can natively integrate with cloud environments, which is important for scalability, resilience, and remote accessibility.

- **Extendable architecture**: Tools with an extendable or plugin architecture can be more easily customized to meet specific needs, allow for integration with other tools, and can evolve as requirements change.

- **Available future roadmap**: Understanding the future development roadmap of a tool can give insights into the longevity and future capabilities of the tool, aligning with long-term strategic goals.

- **Available API**: The presence of a robust and well-documented API is crucial for automation, integration with other tools, and custom development efforts.

Figure 7.3 provides a comparison table that is useful to compare alternative platforms and tools.

Comparison Factor	Importance (1-5)	(C) Capability Rating 0 = Not available or planned, 1 = Capability planned but not available, 2 = Minimal capability available, 3 = Mature capability, 4 = Advanced capability, 5 = Best of breed		
		Alternative1	Alternative2	Alternative3
Compatibility and Integration				
Scope and coverage				
Usability and learning curve				
Automation capabilities				
Reporting and analytics				
Community and support				
Costs				
Performance and scalability				
Security and compliance				
Flexibility and customizability				
Vendor stability and roadmap				
Maintenance				
Feedback mechanisms				
Cloud Native				
Extendable architecture				
Future roadmap				
API				
		0.0	0.0	0.0
		Comparison Score Sum (IxC)		

Figure 7.3 – Comparison of platforms and tools

Remember, these factors can vary in importance based on the specific context and needs of your organization. It's often beneficial to trial a tool before committing to ensure it meets your expectations and requirements.

Example tool platforms and tools

Figures 7.4, *7.5*, and *7.6* provide a list of categories and example tool platforms and tools that can be used for applications of continuous testing, quality, security, and feedback. There are many categories and examples to choose from.

The following figure lists tools starting with the letters A to C.

Tool Platform or Tool Category	Description	Example Tool Platforms or Tools	Tool Currently Used	Rate Current Tool (1-10)
A/B test tools	Compare two versions to determine which one performs better.	Optimizely, Google Optimize, VWO		
Accessibility testing tools	Ensure your app is usable by people with disabilities.	Axe, Wave, Tenon.io		
API test tools	Testing application programming interfaces (APIs).	Postman, SoapUI, Swagger UI		
Application performance monitoring (APM)	Monitor and manage the performance and availability of software.	New Relic, Datadog, Dynatrace		
Application release orhestration/automation.	Orchestrate and automate CI/CD pipelines.	Jenkins, Bamboo, Octopus Deploy, Harness, CloudBees, GitLab		
Artifact repository	Store and manage software images.	Artifactory, Nexus		
Build automation	Automate the process of converting code into a software product.	Maven, Gradle, Ant		
Canary test tools	Roll out changes incrementally, ensure stability before full release.	Istio, Flagger, Spinnaker		
Chaos engineering Tools	Test system robustness by introducing controlled chaos.	Gremlin, Chaos Monkey, Chaos Toolkit		
CI/CD pipeline tools	Automate integration, delivery or deployment of code.	GitLab CI, CircleCI, Travis CI, Harness, GitHub		
Cloud services	Scalable and flexible computing resources, offered over the internet.	AWS, Google Cloud, Azure		
Code collaboration	Enable developers to work together on code in real time.	GitHub, GitLab, Bitbucket		
Code review tools	Facilitate the peer review of code changes.	Review Board, Phabricator, Gerrit		
Code security tools	Detect and fix security vulnerabilities in code.	Sonatype Nexus, Snyk, Veracode		
Communication between teams	Facilitate communication and collaboration among team members.	Slack, Microsoft Teams, Zoom		
Configuration management tools	Automate the provisioning and management of software.	Ansible, Chef, Puppet		
Containers and orchestration	Manage and orchestrate containers across multiple environments.	Docker, Kubernetes, Amazon ECS		
Containers repository	Store and manage container images.	Docker Hub, GitHub Container Registry, GitLab Container Registry		

Figure 7.4 – Example tool platforms and tools (A-C)

The following figure lists tools starting with the alphabet D to P.

Tool Platform or Tool Category	Description	Example Tool Platforms or Tools	Tool Currently Used	Rate Current Tool (1-10)
DAST	Dynamic application security testing.	OWASP ZAP, Acunetix, HCL AppScan		
Data management tools	Assist in the organization, storage, and retrieval of data.	Apache Hadoop, Talend, Informatica		
Database monitoring tools	Monitor database performance and health.	SolarWinds DB Perf. Analyzer, Redgate SQL Monitor, Dynatrace		
Database pipeline tools	Automate database changes and deployments.	Liquibase, Flyway, DbUp		
Database tools	Database development, management, and administration.	MySQL Workbench, MongoDB Compass, Oracle SQL Developer		
Deployment	Assist in deploying applications to various environments.	Heroku, Netlify, Vercel, Octopus Deploy, Harness		
DevSecOps dashboard	Integrate security at every stage of software development.	Jenkins, CircleCI, Travis CI		
Error monitoring and bug tracking	Capture and manage application errors and bugs.	Sentry, Rollbar, Bugsnag		
Feature flag management tools	Manage feature toggles for controlled rollouts.	LaunchDarkly, Split.io, ConfigCat		
Green/Blue deployment tools	Facilitate green/blue deployments.	Spinnaker, AWS CodeDeploy, Azure DevOps		
IAST tools	Interactive application security testing.	Checkmarx, Contrast Security,		
Incident management tools	Manage and respond to IT incidents.	PagerDuty, OpsGenie, VictorOps		
Infrastructure as code tools	Tools that manage and provision infrastructure through code.	Terraform, CloudFormation, Pulumi		
Infrastructure monitoring tools	Monitor the health and performance of physical or virtual servers.	Nagios, Zabbix, Prometheus		
License Compliance Tools	Manage open-source licenses and compliance.	FOSSA, WhiteSource, Black Duck		
Load and performance testing	Assess a system's behavior under controlled load conditions.	JMeter, LoadRunner, Gatling		
Mobile device testing tools	Test applications on various mobile devices.	BrowserStack, Sauce Labs, Appium		
Performance profiling tools	Analyze and optimize application performance.	Blackfire.io, New Relic APM, Dynatrace		
Project management tools	Coordinate team's work and manage projects.	Asana, Trello, Monday.com		

Figure 7.5 – Example tool platforms and tools (D-P)

The following figure lists tools starting with the alphabet Q to Z.

Tool Platform or Tool Category	Description	Example Tool Platforms or Tools	Tool Currently Used	Rate Current Tool (1-10)
RASP tools	Runtime application self-protection.	Prevoty, Signal Sciences, Imperva RASP		
SAST Tools	Static application security testing.	Checkmarx, Fortify, Veracode		
Security scanning and monitoring tools	Scan and monitor for security vulnerabilities.	SonarQube, Fortify, Checkmarx		
Security secrets management tools	Manage and protect secrets like passwords, tokens, and keys.	HashiCorp Vault, AWS Secrets Manager, Azure Key Vault		
Service virtualization Tools	Simulate service behavior for testing purposes.	WireMock, Mountebank, ServiceV Pro		
Software composition analysis tools (SCA)	Analyze and manage open-source components.	Black Duck, WhiteSource, Snyk		
Software IDEs	Integrated development environments for software programmers.	Visual Studio Code, IntelliJ IDEA, Eclipse		
Source code version management	Track and manage changes to software code.	Git, Subversion, Mercurial, Bitbucket		
Static code analysis	Examine source code to find errors or security vulnerabilities.	SonarQube, ESLint, Coverity		
Test automation framework	Manages test process orchestation and automation.	Testsigma, Selenium, Appium, Cucumber, and Robot Framework.		
Test creation tools	Software that assists in the generation of test cases and data.	Selenium IDE, Katalon Studio, TestComplete		
Test environment orchestration	Tools that manage and provision test environments.	Kubernetes, Docker, Terraform		
Test management	Organize and manage the testing process and artifacts.	TestRail, Zephyr, qTest		
Ticket System	Track and manage customer support tickets.	Zendesk, Freshdesk, Jira Service Desk		
Unit test tools	Tools designed to test individual units of source code.	JUnit, NUnit, TestNG		
User analytics and feedback tools	Gain insights from user behavior and feedback.	Google Analytics, Hotjar, UserVoice		
User interface test automation	Automate the testing of a software's user interface.	Selenium WebDriver, Appium, Cypress		
Value stream management platform (VSMP)	Oversee software delivery pipeline and visualize work flow.	Jira, GitLab, Plutora, ConnectAll, CloudBees		
Virtual machines	Emulates a computer system to run	VMware, VirtualBox, Hyper-V		

Figure 7.6 – Example tool platforms and tools (Q-Z)

This section explained the factors that are important to consider when selecting tool platforms and tools for continuous testing, quality, security, and feedback. It provided a useful comparison table for comparing alternative platforms and tools and a categorization for applications of continuous testing, quality, security, and feedback. The next section will describe a methodology for selecting tool platforms and tools.

Methodology for selecting tool platforms and tools

Figure 7.7 illustrates a generally recommended methodology, based on industry best practices, for selecting tools for continuous testing, quality, security, and feedback.

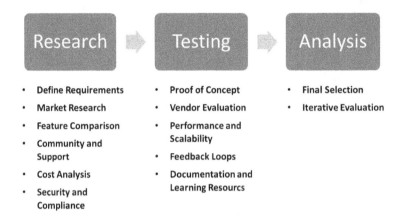

Figure 7.7 – Selection methodology for tool platforms and tools

Here are the activities for each step in the selection process:

1. **Define your requirements**: Start by understanding the specific needs of your organization. This includes the scale of operations, the complexity of projects, the skills of your team, and the specific goals you want to achieve with continuous testing, quality, security, and feedback.

2. **Market research**: Conduct research to identify the tools available in the market that meet your needs. This should include both open-source and commercial options. *Figure 7.2* illustrates factors to consider when choosing between source options.

3. **Feature comparison**: Compare the features of the different tools against your requirements. Pay particular attention to aspects such as cloud-native capabilities, extendable architecture, available future roadmap, and available APIs. *Figure 7.3* illustrates an example of a comparison chart that can be used for organizing the comparison.

4. **Community and support**: Evaluate the community support for open-source tools and customer service for commercial products. A strong community or good customer support can be crucial for solving problems that may arise.

5. **Cost analysis**: Consider the total cost of ownership, including licensing fees for vendor products, operational costs for open-source tools, and the development and maintenance costs for homegrown solutions.

6. **Security and compliance**: Verify that the tools comply with the relevant security and data protection regulations that your organization is subject to.

7. **Proof of concept**: Implement a proof of concept for the shortlisted tools to see how they perform in your environment. This will also help assess the learning curve and ease of integration with your existing systems.

8. **Vendor evaluation**: If considering vendor tools, evaluate the vendor's stability, reputation, and the robustness of their product roadmap.

9. **Performance and scalability**: Ensure that the tools perform well under expected loads and can scale with your business needs.

10. **Integration capabilities**: The tools should be able to integrate seamlessly with your existing CI/CD pipeline and other tools.

11. **Feedback loops**: Tools should facilitate quick feedback loops. Tools focusing on feedback should allow for easy gathering, analysis, and implementation of user and stakeholder feedback.

12. **Documentation and learning resources**: Ensure that there are comprehensive documentation and learning resources available to enable your team to get up to speed quickly.

13. **Final selection**: Use a scoring system to evaluate how each tool stacks up against your criteria. Involve key stakeholders in the final decision to ensure buy-in.

14. **Iterative evaluation**: Tools should be evaluated regularly to ensure they continue to meet the needs of the organization as it grows and evolves.

Remember, the selection of tools is a strategic decision that can have a long-term impact on your organization's ability to deliver quality software efficiently and securely. Therefore, it is essential to invest the necessary time and resources to make an informed decision.

Determining how many tools are enough

Selecting the right number of tool platforms and tools for continuous testing, quality, security, and feedback within an enterprise is influenced by several key factors. The goal is to achieve a balance that provides comprehensive coverage and integration capabilities without causing tool sprawl or unnecessary complexity. Here's what typically determines how many tools are enough:

- **Scope of projects**: The size, number, and complexity of projects undertaken by the enterprise dictate the range of tools needed. Larger projects with more complex architectures may require specialized tools for different aspects or layers of testing and security.

- **Technology stack**: The diversity of the technology stack used in the enterprise's applications can necessitate different tools. Each layer (frontend, backend, database, etc.) and technology (languages, frameworks) might have specific tooling for optimal testing and security.

- **Integration capabilities**: Tools should seamlessly integrate with existing workflows, CI/CD pipelines, and other tools. The degree of integration required can limit or expand the number of tools that can be effectively utilized without creating silos or integration headaches.

- **Compliance and regulatory requirements**: Depending on the industry, enterprises may need to comply with specific regulatory standards (e.g., GDPR, HIPAA, PCI-DSS) that dictate certain levels of security, quality, and feedback mechanisms, influencing the choice and number of tools.

- **Skillset of the team**: The expertise and experience of the development, QA, and security teams with certain tools or types of tools can influence the selection. It's essential to choose tools that the team can effectively use and manage.

- **Budget constraints**: The cost of tooling, including licenses, training, and maintenance, plays a significant role. It's about finding the right mix of tools that offer the necessary capabilities within the available budget.

- **Overall strategy and goals**: The enterprise's strategy regarding speed to market, quality benchmarks, security posture, and customer satisfaction goals will heavily influence how comprehensive the tooling needs to be.

- **Feedback and continuous improvement**: The effectiveness of tools in providing actionable feedback and allowing for continuous improvement is crucial. Tools that offer comprehensive insights without overwhelming the team are preferred.

Balancing act

The right number of tools provides thorough coverage across testing, quality, security, and feedback without overlapping functionalities that could waste resources or create confusion. It's about achieving a balance where each tool adds distinct value, integrates well with others, and supports the enterprise's overall objectives without creating an unwieldy toolchain that slows down processes or complicates workflows.

Regularly evaluating the toolchain's effectiveness and adjusting as projects evolve, technologies change, and the enterprise grows is key to maintaining this balance.

Summary

This chapter outlined the significance of tool platforms and tools in enhancing software development through continuous testing, quality, security, and feedback. It elaborated on how these tools and platforms serve distinct yet interconnected roles in automating, integrating, and streamlining development processes. The key to selection is understanding each tool's functionalities and integration capabilities and aligning them with strategic objectives. The chapter further detailed a methodology for choosing the right tools, emphasizing compatibility, automation, and cost among other factors, ensuring they complement the development life cycle and support continuous improvement.

In the next chapter, I will explore **artificial intelligence/machine learning (AI/ML)** applications for continuous testing, quality, security, and feedback.

Applying AL/ML to Continuous Testing, Quality, Security, and Feedback

This chapter delves into the transformative role of **artificial intelligence** (**AI**) and **machine learning** (**ML**) across the software development life cycle, with a special focus on enhancing continuous testing, quality, security, and feedback practices.

The chapter starts with an overview of AI/ML applications. It explains how these technologies are reshaping the landscapes of continuous testing, quality, security, and feedback. Each section provides in-depth insights into AI/ML strategies that are designed to automate and optimize processes, from early-stage code testing to post-deployment monitoring. This is to facilitate a seamless, continuous integration and delivery pipeline with an AI-enabled toolchain.

The chapter prescribes a methodology for selecting AI/ML-enabled tools that can integrate effectively within your continuous testing, quality, security, and feedback transformation projects.

By the end of this chapter, you will have gained a comprehensive understanding of AI/ML-enabled tools that are useful for continuous testing, quality, security, and feedback. You will have also learned a systematic approach to selecting AI/ML-enabled tools.

In this chapter, we'll cover the following main topics:

- AI/ML applications
- AI/ML for continuous testing
- AI/ML for continuous quality
- AI/ML for continuous security
- AI/ML for continuous feedback
- Methodology for selecting AI/ML tools

Let's get started!

AI/ML applications

In the rapidly evolving landscape of software development and operations, AI and ML are transforming the way organizations approach continuous testing, quality, security, and feedback.

As illustrated in *Figure 8.1*, these technologies have become pivotal in enabling organizations to navigate the complexities of digital transformation, especially within frameworks such as DevOps, DevSecOps, and SRE. By harnessing the power of AI/ML, businesses are not only accelerating their development cycles but are also enhancing the robustness and security of their applications in unprecedented ways.

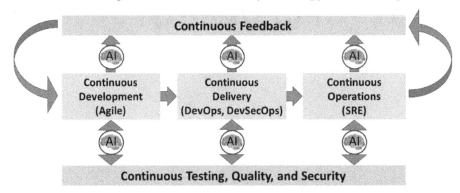

Figure 8.1 – AI/ML applications for continuous testing, quality, security, and feedback

AI/ML technologies have significantly advanced in recent years, reaching a level of sophistication that allows for their effective integration into various stages of the **continuous integration/continuous deployment (CI/CD)** pipelines, where the need for speed and efficiency must be balanced with the demands for quality and security. AI/ML contributes by automating complex tasks, predicting potential issues before they occur, and providing actionable insights, thereby reducing manual efforts and enabling more strategic use of human resources.

The application of AI/ML encompasses the capability to learn from data, adapt to new information, and improve over time. This ability is invaluable for identifying patterns, anticipating vulnerabilities, and optimizing testing strategies. In quality assurance, AI-driven tools can predict areas most likely to fail and tailor testing efforts accordingly. In security, ML algorithms can detect anomalies that signify potential threats, while in operations, AI can enhance feedback mechanisms, leading to more resilient and responsive systems.

The subsequent sections of this chapter will delve into specific use cases of AI/ML within the realms of continuous testing, quality, security, and feedback. These examples will illustrate how AI/ML not only streamlines processes but also elevates the standards of software development and maintenance. This chapter provides a comprehensive overview of AI/ML's transformative potential for organizations committed to excellence in their digital transformation journeys.

AI/ML for continuous testing

Integrating AI and ML into the software's continuous testing activities can significantly streamline processes and address potential bottlenecks in each activity, as illustrated in *Figure 8.2*.

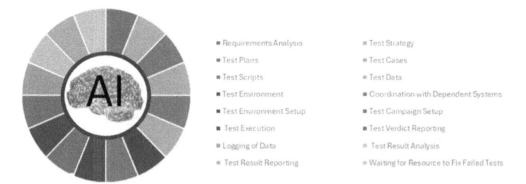

- Requirements Analysis
- Test Plans
- Test Scripts
- Test Environment
- Test Environment Setup
- Test Execution
- Logging of Data
- Test Result Reporting
- Test Strategy
- Test Cases
- Test Data
- Coordination with Dependent Systems
- Test Campaign Setup
- Test Verdict Reporting
- Test Result Analysis
- Waiting for Resource to Fix Failed Tests

Figure 8.2 – AI/ML for continuous testing activities

Here's how these technologies can be applied across various testing activities:

1. **Requirements analysis**:

    -   *Explanation*: Ensures that test scenarios align with business requirements and user needs.

    -   *Bottleneck*: Misinterpretation or incomplete analysis can lead to inadequate test coverage.

    -   *AI/ML solution*: NLP can automate the extraction and interpretation of requirements, ensuring comprehensive and accurate test coverage.

2. **Test strategy**:

    -   *Explanation*: Outlines the testing approach, objectives, and resources.

    -   *Bottleneck*: An unclear strategy may lead to inefficient testing efforts and resource allocation.

    -   *AI/ML solution*: AI can analyze historical data to suggest the most effective test strategies and predict resource needs.

3. **Test plans**:

    -   *Explanation*: Detailed documents guiding the testing process, timelines, and responsibilities.

    -   *Bottleneck*: Inflexible plans can struggle to adapt to project changes, causing delays.

    -   *AI/ML solution*: ML algorithms can suggest adjustments to test plans based on ongoing project developments and past outcomes.

4. **Test cases**:

   - *Explanation*: Specific conditions under which a test is executed.

   - *Bottleneck*: Time-consuming development and significant effort to maintain test cases.

   - *AI/ML solution*: AI can automate the generation of test cases from requirements documents, improving efficiency and coverage.

5. **Test scripts**:

   - *Explanation*: Automated scripts that execute test cases.

   - *Bottleneck*: Script development and maintenance can be resource-intensive.

   - *AI/ML solution*: AI can generate and update test scripts based on changes in the application or test cases, reducing maintenance effort.

6. **Test data**:

   - *Explanation*: Datasets used during testing to simulate real-world scenarios.

   - *Bottleneck*: Creating, managing, and maintaining accurate test data is challenging.

   - *AI/ML solution*: AI can automate the generation and management of test data, ensuring relevance and variety.

7. **Test environment**:

   - *Explanation*: The setup where testing is conducted, mirroring production environments as closely as possible.

   - *Bottleneck*: Configuration and maintenance of test environments are complex.

   - *AI/ML solution*: AI can predict and configure optimal test environments based on test requirements, reducing setup time.

8. **Coordination with dependent systems**:

   - *Explanation*: Ensuring the system under test interacts correctly with databases and other applications.

   - *Bottleneck*: Dependency management can cause delays.

   - *AI/ML solution*: AI can automate the detection and resolution of integration issues, enhancing coordination efficiency.

9. **Test campaign setup**:

   - *Explanation*: Organizing and scheduling a series of test executions.

   - *Bottleneck*: Requires careful planning and can be hindered by resource limitations.

   - *AI/ML solution*: AI can assist in scheduling and prioritizing test campaigns based on risk and impact analysis.

10. **Test execution**:

    - *Explanation*: The process of running test cases and scripts, both automated and manual.

    - *Bottleneck*: Time-consuming, particularly for manual tests.

    - *AI/ML solution*: AI can prioritize test execution and identify flaky tests, streamlining the process.

11. **Test verdict reporting**:

    - *Explanation*: Determining and reporting the outcome of test executions.

    - *Bottleneck*: Manual verdict determination can be slow.

    - *AI/ML solution*: AI can automatically interpret test outcomes, speeding up reporting.

12. **Logging of data**:

    - *Explanation*: Recording data relevant to the test for further analysis.

    - *Bottleneck*: Extensive data collection can overwhelm resources.

    - *AI/ML solution*: AI can intelligently filter and log pertinent data, reducing noise.

13. **Test result analysis**:

    - *Explanation*: Analyzing test outcomes to identify defects and issues.

    - *Bottleneck*: Requires significant time and expertise.

    - *AI/ML solution*: ML algorithms can quickly identify patterns and anomalies in test results, highlighting potential issues.

14. **Test result reporting**:

    - *Explanation*: Communicating findings to stakeholders.

    - *Bottleneck*: Compiling reports is time-intensive.

    - *AI/ML solution*: Automated reporting tools powered by AI can generate insightful and comprehensive reports quickly.

15. **Waiting for resources to fix failed tests**:

- *Explanation*: Downtime while awaiting fixes for identified issues.

- *Bottleneck:* Halts testing progress.

- *AI/ML solution:* AI can predict which areas might fail and propose potential fixes.

Applying AI/ML to software testing activities offers numerous advantages but also introduces several challenges, as illustrated in *Figure 8.3*. Addressing these problems requires a combination of technical solutions, process adjustments, and cultural changes.

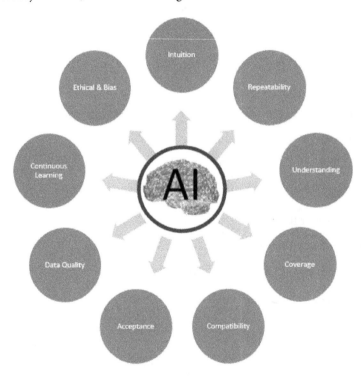

Figure 8.3 – AI/ML challenges for continuous testing

Here, we discuss some common issues and strategies for overcoming them:

- **Lack of intuitive understanding of the application being tested**:

  - *Problem*: AI/ML models may not fully grasp the application's context or the nuances of its functionalities, leading to less effective test scenarios.

  - *Solution*: Enhance AI models with richer contextual data and incorporate feedback loops where testers can refine and adjust AI-generated test cases. Employing techniques such as reinforcement learning can also help AI models better understand application contexts over time.

- **Repeatability and consistency between test sessions**:

  - *Problem*: AI-driven tests may generate different outputs for the same input over different sessions, complicating test consistency and traceability.

  - *Solution*: Implement versioning for AI models and their training data, ensuring consistency across test sessions. Use deterministic approaches in conjunction with AI to maintain a core of stable, repeatable tests.

- **Lack of understanding of the tests generated**:

  - *Problem*: Testers may find it challenging to understand or trust the rationale behind AI-generated test cases, impacting their ability to evaluate test outcomes effectively.

  - *Solution*: Incorporate explainability features into AI/ML models to provide insights into their decision-making processes. Foster a culture of trust and understanding through education and transparency regarding how AI models operate.

- **Test coverage**:

  - *Problem*: There's a risk that AI/ML might not adequately cover all testing scenarios, potentially missing critical defects.

  - *Solution*: Combine AI/ML with traditional testing methods to ensure comprehensive coverage. Regularly review and adjust the criteria used by AI/ML models to generate test cases, ensuring they align with evolving application features and risks.

- **Compatibility with different test tools**:

  - *Problem*: AI/ML models might not seamlessly integrate with existing testing tools and frameworks, limiting their utility.

  - *Solution*: Develop or use AI/ML solutions with extensive API support and integration capabilities. Work with tool vendors or contribute to open source projects to enhance compatibility.

- **Acceptance by teams**:

  - *Problem*: Testers and developers may be skeptical or resistant to AI-driven testing due to concerns about job displacement or mistrust of AI's effectiveness.

  - *Solution*: Educate and involve teams in the development and implementation of AI/ML testing strategies. Demonstrate the value of AI/ML in augmenting their roles rather than replacing them, focusing on AI as a tool to tackle mundane tasks and allowing them to focus on more complex and rewarding work.

- **Data quality and availability**:

  - *Problem*: AI/ML models require large amounts of high-quality data for training. Inadequate or poor-quality data can lead to ineffective testing.

  - *Solution*: Invest in data curation and generation strategies, such as synthetic data creation, to ensure that models are well trained.

- **Continuous learning and adaptation**:

  - *Problem*: AI/ML models may become outdated as applications evolve.

  - *Solution*: Establish continuous learning mechanisms where models are regularly updated with new data and feedback, ensuring that they remain relevant and effective.

- **Ethical and bias considerations**:

  - *Problem*: AI/ML testing models might inherit or amplify biases present in their training data, leading to unfair or discriminatory outcomes.

  - *Solution*: Implement ethical guidelines and bias detection methodologies for AI/ML model development and use. Regularly audit models for biases and correct them as needed.

By addressing these challenges with thoughtful strategies, organizations can maximize the benefits of AI/ML in testing activities while mitigating potential drawbacks, leading to more efficient, effective, and trustworthy testing processes.

Real-world use case for AI/ML-assisted continuous testing

A real-world application is seen in the use of a tool such as **Applitools**, which utilizes visual AI to automate and streamline the validation of **user interfaces** (**UIs**) across multiple web and mobile applications. This AI-driven approach enables teams to detect discrepancies or visual regressions by comparing the current version of the application's UI against baseline images that have been previously captured and verified as correct.

This method drastically reduces the time and effort required for manual testing by automating the detection of visual issues such as layout problems, color mismatches, or unexpected changes in UI elements. The AI component adapts to changes in the UI over time, thereby maintaining its effectiveness even as the application evolves. By integrating such tools into the development pipeline, organizations can ensure more accurate, efficient, and scalable testing processes, ultimately leading to faster deployment cycles and higher-quality software products.

AI/ML for continuous quality

Implementing continuous quality across the development, delivery, and production life cycle involves several activities designed to ensure stable releases and enhance user satisfaction, as illustrated in *Figure 8.4.*

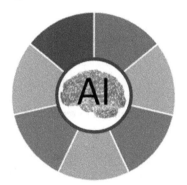

- Quality Metrics Integration
- Continuous Testing
- Predictive Bug and Issue Detection
- Production Monitoring and Anomaly Detection

- Automated Code Reviews
- Real-Time User Feedback Analysis
- Deployment Risk Assessment

Figure 8.4 – AI/ML for continuous quality activities

Here is a list of activities essential for this approach, along with potential bottlenecks and how AI/ML can address these challenges:

1. **Quality metrics integration**:

    - *Description*: Embedding quality metrics into every phase of the software development life cycle to monitor and improve quality continuously.

    - *Bottlenecks*: Manual collection and analysis of quality metrics can be time-consuming and prone to errors, potentially slowing down the development process.

    - *AI/ML application*: AI can automate the extraction, monitoring, and analysis of quality metrics from various tools and platforms, providing real-time insights and predictions to prevent quality issues.

2. **Automated code reviews**:

    - *Description*: Utilizing tools to automatically review code for potential issues, adherence to coding standards, and security vulnerabilities as soon as it is committed.

    - *Bottlenecks*: High false positive rates in automated code review tools can lead to developer fatigue and slow down the review process.

    - *AI/ML application*: ML models can learn from historical code review data to reduce false positives and highlight the most relevant issues, streamlining the review process.

3. **Continuous testing**:

   - *Description*: Running automated tests as part of the CI/CD pipeline to identify defects as early as possible.

   - *Bottlenecks*: Creating and maintaining a comprehensive test suite that covers all aspects of the application can be resource-intensive and slow down releases.

   - *AI/ML application*: AI can assist in generating test cases based on changes in the code base and user behavior, ensuring relevant and efficient test coverage.

4. **Real-time user feedback analysis**:

   - *Description*: Collecting and analyzing user feedback from various channels in real time to identify issues and areas for improvement.

   - *Bottlenecks*: Manual analysis of user feedback from multiple sources can be overwhelming and delay the identification of critical issues.

   - *AI/ML application*: NLP and sentiment analysis algorithms can automatically categorize and prioritize user feedback, enabling faster response to critical issues and trends.

5. **Predictive bug and issue detection**:

   - *Description*: Predicting potential bugs and issues before they occur based on historical data and patterns in code changes.

   - *Bottlenecks*: Identifying potential issues before they manifest can be challenging without historical context, possibly leading to unnoticed problems until after release.

   - *AI/ML application*: ML models can analyze code changes, commit history, and issue trackers to predict areas of the code base most likely to introduce defects, allowing preemptive action.

6. **Deployment risk assessment**:

   - *Description*: Assessing the risk associated with a new release based on quality metrics, test results, and historical deployment data.

   - *Bottlenecks*: Manual risk assessments can be subjective and inconsistent, potentially leading to unnecessary delays or overlooked issues.

   - *AI/ML application*: AI algorithms can provide objective risk assessments by analyzing extensive datasets, helping teams make informed decisions about releases.

7. **Production monitoring and anomaly detection**:

   - *Description*: Monitoring production environments for unexpected behavior, performance issues, and security threats.

- *Bottlenecks*: Sifting through vast amounts of monitoring data to identify anomalies can delay the detection and resolution of issues.

- *AI/ML application*: ML models can continuously analyze monitoring data to detect anomalies in real time, reducing detection time and improving response efficiency.

By integrating these activities into the development, delivery, and production life cycle, organizations can significantly enhance their continuous quality strategy. AI/ML applications play a crucial role in overcoming bottlenecks associated with these activities, enabling more stable releases and higher user satisfaction.

Applying AI/ML to continuous quality activities introduces significant advantages, yet it also comes with challenges that can impede its effectiveness, as illustrated in *Figure 8.5*.

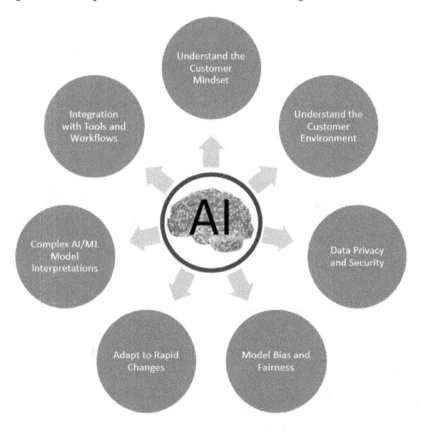

Figure 8.5 – AI/ML challenges for continuous quality

Recognizing these problems is crucial for developing strategies to mitigate them. Here are some common issues associated with integrating AI/ML into continuous quality efforts, along with proposed solutions:

- **Lack of intuitive understanding of the customer's mindset**:

  - *Problem*: AI/ML models may not inherently grasp the nuances of customer expectations or how users interact with the application, potentially leading to misaligned quality improvements.

  - *Strategy*: To bridge this gap, combine AI/ML insights with direct customer feedback mechanisms and user behavior analysis. Utilizing NLP to analyze customer reviews and feedback can provide qualitative insights that inform model training and adjustment, aligning AI-driven quality improvements with user expectations.

- **Uncertainty about the customer's environment**:

  - *Problem*: AI/ML models may struggle to predict and test for the vast array of user environments (devices, operating systems, and network conditions), potentially missing critical quality issues.

  - *Strategy*: Implementing synthetic data generation and simulation techniques can help create diverse scenarios that mimic a wide range of customer environments. This, coupled with real-world usage data, can train AI/ML models to better anticipate and address environment-specific quality issues.

- **Data privacy and security concerns**:

  - *Problem*: Collecting and utilizing data for AI/ML, especially user feedback and behavior data, raises concerns about privacy and data security.

  - *Strategy*: Employ privacy-preserving data analysis techniques, such as differential privacy and federated learning, to train models without compromising individual user data. Ensure compliance with data protection regulations by adopting a data governance framework that prioritizes user consent and data minimization.

- **Model bias and fairness**:

  - *Problem*: AI/ML models may inadvertently learn biases present in their training data, leading to unfair or discriminatory outcomes in quality improvements.

  - *Strategy*: Regularly audit AI/ML models for bias and implement fairness-aware ML practices. This involves diversifying training data, applying debiasing techniques, and setting fairness criteria to evaluate model outputs.

- **Adaptability to rapid changes**:

  - *Problem*: AI/ML models trained on historical data may not quickly adapt to rapid changes in user behavior, market trends, or the introduction of new features.

  - *Strategy*: Incorporate continuous learning mechanisms into AI/ML models to allow for frequent retraining and updates based on new data. Leveraging techniques such as online learning can enable models to adapt to changes in real time.

- **Complexity of AI/ML model interpretability**:

  - *Problem*: The "black box" nature of some AI/ML models can make it challenging for teams to understand and trust their predictions and recommendations, especially regarding quality improvements.

  - *Strategy*: Focus on developing and employing **explainable AI (XAI)** methods that provide insight into the decision-making process of AI models. Foster a culture of transparency by offering training and resources that help team members understand how AI/ML models contribute to quality outcomes.

- **Integration with existing tools and workflows**:

  - *Problem*: Seamlessly integrating AI/ML solutions into existing continuous quality processes and tools can be challenging, potentially leading to disruptions and inefficiencies.

  - *Strategy*: Adopt AI/ML tools that offer extensive API support, plugins, and integration capabilities with existing quality assurance and development platforms. Consider incremental integration strategies that allow for gradual adaptation and learning.

Addressing these challenges requires a thoughtful approach that combines technical solutions, process adjustments, and continuous learning. By acknowledging and strategically tackling these issues, organizations can fully harness the potential of AI/ML to enhance continuous quality initiatives, leading to improved user satisfaction and reduced production failure rates.

Real-world use case for AI/ML-assisted continuous quality

A practical example of using AI/ML-assisted tools for maintaining continuous quality in software development is seen with **SonarQube**, an open-source platform that leverages ML to enhance code quality analysis. SonarQube scans code bases for bugs, vulnerabilities, and code smells by employing static analysis methods enhanced by ML algorithms. These algorithms learn from vast datasets of code to better identify complex coding issues that traditional methods might miss.

This ML capability enables SonarQube to adaptively improve its analysis accuracy over time, learning from the patterns and corrections made in various code bases. By integrating SonarQube into the CI/CD pipeline, developers receive real-time feedback on code quality, ensuring that quality checks are an integral part of the development process rather than an afterthought. This continual, automated review helps maintain high coding standards, reducing both the technical debt and the potential for defects in the production environment.

AI/ML for continuous security

Implementing continuous security involves integrating proactive and reactive security measures seamlessly across the development, delivery, and operations life cycle. This approach is designed to minimize the frequency and impact of security events. *Figure 8.6* illustrates key activities needed to achieve continuous security.

- Security Requirements Analysis
- Automated Vulnerability Scanning
- Threat Modelling and Risk Assessment
- Incident Response and Remediation
- Secure Coding Practices Training
- Dynamic Application Security Testing (DAST)
- Security Incident Detection
- Continuous Feedback Loop

Figure 8.6 – AI/ML for continuous security activities

The following explains potential bottlenecks and how AI/ML can mitigate these challenges:

1. **Security requirements analysis**:

   - *Description*: Defining and understanding the security requirements specific to the application and the environment it operates in.

   - *Bottlenecks*: Time-consuming analysis and potential for overlooking critical requirements.

   - *AI/ML application*: AI-powered tools can analyze project documentation and code to automatically identify security requirements and regulations applicable to the project, speeding up the process and reducing oversights.

2. **Secure coding practices training**:

   - *Description*: Training development teams in secure coding practices to prevent introducing vulnerabilities.

   - *Bottlenecks*: Keeping training materials up to date and ensuring all developers have the latest knowledge can be challenging.

   - *AI/ML application*: ML algorithms can curate personalized training content based on the most common security mistakes identified in the code base, ensuring relevant and timely training.

3. **Automated vulnerability scanning**:

   - *Description*: Regularly scanning the code base and dependencies for known vulnerabilities using automated tools.

   - *Bottlenecks*: High false positive rates can overwhelm developers, and scanning can slow down the CI/CD pipeline.

   - *AI/ML application*: AI models can prioritize vulnerabilities based on historical data, reducing the noise of false positives and focusing efforts on the most critical issues.

4. **Dynamic application security testing (DAST)**:

   - *Description*: Conducting automated security testing on running applications to identify runtime vulnerabilities.

   - *Bottlenecks*: DAST can be resource-intensive and slow, potentially delaying deployments.

   - *AI/ML application*: AI can optimize test runs by focusing on areas with recent changes or known vulnerabilities, improving speed and efficiency.

5. **Threat modeling and risk assessment**:

   - *Description*: Analyzing potential threats and assessing risks to prioritize security efforts.

   - *Bottlenecks*: Manual threat modeling is time-consuming and may not capture the evolving threat landscape.

   - *AI/ML application*: AI algorithms can automate threat modeling by analyzing code changes and external threat intelligence, providing real-time risk assessments.

6. **Security incident detection**:

   - *Description*: Monitoring applications and infrastructure for security incidents using automated tools.

   - *Bottlenecks*: The volume of alerts can overwhelm security teams, leading to missed or ignored threats.

   - *AI/ML application*: ML can enhance anomaly detection, distinguishing between normal behavior and potential security incidents, reducing false positives and alert fatigue.

7. **Incident response and remediation**:

   - *Description*: Responding to and mitigating the impact of security incidents quickly and efficiently.

   - *Bottlenecks*: Manual response processes can be slow, increasing the time to resolution.

   - *AI/ML application*: AI-driven automation can trigger predefined response actions for common incidents, speeding up resolution times and freeing up resources for more complex analysis.

8. **Continuous feedback loop**:

- *Description*: Incorporating feedback from security operations back into development to prevent future incidents.

- *Bottlenecks*: Siloed teams and processes can hinder the effective communication of feedback.

- *AI/ML application*: AI tools can analyze incident reports and feedback to identify patterns and recommend changes to development practices, fostering a culture of continuous improvement.

By addressing each of these activities with the support of AI/ML applications, organizations can significantly enhance their continuous security posture, ensuring that security measures evolve in tandem with development practices and emerging threats.

Applying AI/ML to continuous security activities offers significant benefits, such as automating repetitive tasks and enhancing detection capabilities. However, this integration comes with its own set of challenges, as illustrated in *Figure 8.7*.

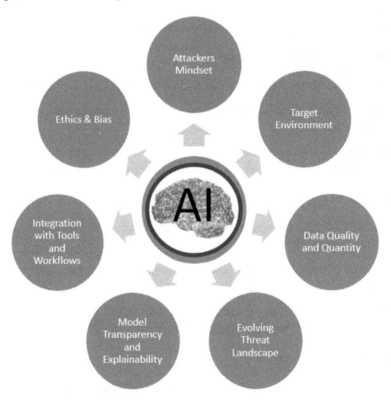

Figure 8.7 – AI/ML challenges for continuous security

Addressing these issues is crucial for leveraging AI/ML effectively within a continuous security framework. Here are common problems and strategies to overcome them:

- **Lack of intuitive understanding of the attacker's mindset**:

  - *Problem*: AI/ML models may not fully capture the creativity and adaptability of human attackers, potentially missing novel or sophisticated attack vectors.

  - *Strategy*: Incorporate adversarial AI techniques and red team exercises to train AI/ML models on a broader spectrum of attack scenarios, including those that require human-like intuition and creativity. Engage in continuous learning practices where AI/ML systems are regularly updated with insights from the latest threat intelligence and real-world attack patterns.

- **Uncertainty about the target environment**:

  - *Problem*: AI/ML systems might not have complete knowledge of the target environment, leading to inaccurate threat modeling and vulnerability assessments.

  - *Strategy*: Use AI/ML in conjunction with dynamic discovery tools to continuously update the system's understanding of the environment. Implement hybrid models that combine ML insights with input from security experts to ensure comprehensive coverage of environmental variables.

- **Data quality and quantity**:

  - *Problem*: AI/ML models require large volumes of high-quality data to train effectively. Insufficient or poor-quality data can lead to inaccurate predictions and models.

  - *Strategy*: Leverage synthetic data generation techniques to augment training datasets, ensuring AI/ML models have access to diverse and comprehensive data. Establish partnerships and data-sharing agreements with trusted entities to enrich the dataset further.

- **Evolving threat landscape**:

  - *Problem*: The rapid evolution of the cyber threat landscape can outpace the learning capabilities of AI/ML models, making them less effective over time.

  - *Strategy*: Implement continuous learning mechanisms that allow AI/ML models to adapt to new threats in real time. This includes the integration of automated threat intelligence feeds and the use of unsupervised learning techniques to detect novel patterns.

- **Model transparency and explainability**:

  - *Problem*: The "black box" nature of some AI/ML models can make it difficult for security teams to understand the reasoning behind certain decisions, impacting trust and accountability.

  - *Strategy*: Focus on developing and employing XAI techniques that offer insights into the decision-making process of models. This includes using models that inherently provide more transparency or adopting tools that can interpret model outputs.

- **Integration with existing security tools and workflows**:

  - *Problem*: Seamlessly integrating AI/ML into existing security tools and workflows can be challenging, potentially leading to operational inefficiencies.

  - *Strategy*: Prioritize AI/ML solutions that offer robust API support and compatibility with standard security tools and platforms. Adopt a phased integration approach that allows for gradual adjustment and optimization of workflows.

- **Ethical considerations and bias**:

  - *Problem*: AI/ML models can inherit biases from their training data, potentially leading to unethical outcomes or discriminatory practices in security operations.

  - *Strategy*: Conduct regular audits of AI/ML models to identify and mitigate biases. Incorporate diverse datasets in the training phase and engage multidisciplinary teams in the development process to ensure that ethical considerations are prioritized.

By acknowledging and strategically addressing these challenges, organizations can more effectively harness the power of AI/ML in their continuous security efforts, leading to enhanced detection capabilities, improved response times, and a more resilient security posture.

Real-world use case for AI/ML-assisted continuous security

A notable real-world application of AI/ML-assisted tools in continuous security is shown through **Darktrace**, an AI-driven cybersecurity platform. Darktrace employs ML algorithms to learn the normal behavior of an organization's network, enabling it to detect and respond to threats in real time. By continuously monitoring network traffic and using unsupervised ML to build an understanding of "self" for every device, user, and network within an organization, Darktrace can identify unusual behaviors that may indicate a cyber-attack.

This proactive approach allows the system to autonomously respond to in-progress cyber threats quickly, often mitigating risks before they escalate into serious breaches. For instance, if Darktrace detects an unknown device attempting to make unusual data transfers, it can automatically interrupt these potentially malicious activities, safeguarding sensitive data effectively. This AI-enhanced monitoring and response capability is a significant advancement over traditional, rule-based security systems, enabling organizations to adapt to the evolving security landscape dynamically.

AI/ML for continuous feedback

Implementing continuous feedback involves systematically gathering, analyzing, and acting on feedback from users and stakeholders throughout the development, delivery, and production life cycle. This process is designed to enhance the software's reliability and the team's responsiveness to changes. *Figure 8.8* illustrates activities essential for continuous feedback.

- Feedback Collection
- Feedback Analysis
- Integration with Development Workflows
- Feature Implementation and Testing
- Release and Monitoring
- Feedback Loop Closure
- Impact Analysis

Figure 8.8 – AI/ML for continuous feedback activities

Potential bottlenecks and AI/ML solutions to address these challenges are explained in the following list:

1. **Feedback collection**:

    - *Description*: Gathering feedback from various sources, including user surveys, support tickets, and social media.

    - *Bottlenecks*: The volume and variety of feedback can overwhelm manual processing efforts, leading to slow response times.

    - *AI/ML application*: NLP and sentiment analysis can automatically categorize and prioritize feedback, helping teams quickly identify and address the most critical issues.

2. **Feedback analysis**:

    - *Description*: Analyzing the collected feedback to identify common themes, user pain points, and potential enhancements.

    - *Bottlenecks*: Manual analysis is time-consuming and may not accurately capture the full spectrum of user sentiments.

    - *AI/ML application*: AI-driven text analytics and pattern recognition can reveal insights from large datasets of feedback, highlighting areas for improvement that might not be immediately obvious.

3. **Integration into development workflow**:

    - *Description*: Incorporating actionable feedback into the development backlog and prioritizing it within sprints.

    - *Bottlenecks*: Integrating feedback into existing development workflows can disrupt planned schedules and resource allocations.

- *AI/ML application*: ML algorithms can assess the impact and effort of implementing feedback, automatically suggesting priorities and adjustments to the development roadmap.

4. **Feature implementation and testing**:

    - *Description*: Developing and testing new features or fixes based on user feedback.

    - *Bottlenecks*: Rapidly implementing and testing changes based on feedback can strain resources and potentially introduce new issues.

    - *AI/ML application*: Automated testing tools powered by AI can quickly validate new features and fixes, ensuring they meet quality standards without significantly slowing down development.

5. **Release and monitoring**:

    - *Description*: Deploying updates to users and monitoring the impact of changes on system reliability and user satisfaction.

    - *Bottlenecks*: Continuous deployment of changes risks destabilizing the production environment and affecting system reliability.

    - *AI/ML application*: AI-based monitoring tools can detect anomalies and regressions in real time, allowing for quick recovery actions and minimizing negative impact on users.

6. **Feedback loop closure**:

    - *Description*: Informing stakeholders and users about the actions taken in response to their feedback, closing the feedback loop.

    - *Bottlenecks*: Effectively communicating back to a large user base about feedback implementation can be challenging and resource-intensive.

    - *AI/ML application*: Automated communication tools, such as chatbots or personalized emails, can inform users about the status of their feedback, enhancing transparency and trust.

7. **Impact analysis**:

    - *Description*: Evaluating the effectiveness of changes made in response to feedback by analyzing metrics related to user satisfaction and system reliability.

    - *Bottlenecks*: Manually correlating feedback-driven changes with outcomes in system performance and user satisfaction can be complex and imprecise.

    - *AI/ML application:* Advanced analytics and ML models can measure the impact of specific changes on key performance indicators, providing clear insights into the value of feedback implementation.

By integrating these activities and leveraging AI/ML applications, organizations can significantly improve their continuous feedback process. This approach not only accelerates the implementation of valuable feedback but also ensures that changes positively affect system reliability and user satisfaction.

Applying AI/ML to continuous feedback processes can transform how organizations collect, analyze, and act on feedback. However, as illustrated in *Figure 8.9*, integrating these technologies comes with its own set of challenges. Understanding these problems is crucial to developing effective strategies to mitigate them.

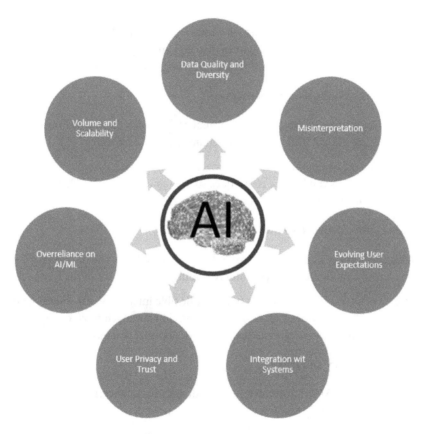

Figure 8.9 – AI/ML challenges for continuous feedback

Here are common issues associated with leveraging AI/ML in continuous feedback activities, along with proposed solutions:

- **Data quality and diversity**:

  - *Problem*: AI/ML models require high-quality, diverse data to accurately analyze feedback. Poor quality or biased data can lead to inaccurate insights and misinformed decisions.

- *Strategy*: Implement robust data collection and preprocessing methods to ensure data quality and representativeness. Regularly review and update data collection strategies to mitigate biases and improve the diversity of the feedback data.

- **Misinterpretation of feedback**:

  - *Problem*: AI/ML models, especially those based on NLP, might misinterpret the nuances and context of user feedback, potentially leading to incorrect prioritization or misunderstanding of user needs.

  - *Strategy*: Combine AI/ML analysis with human review, particularly for feedback that is complex or carries significant weight in decision-making. Employing a hybrid approach ensures that AI-driven insights are validated through human expertise.

- **Adaptability to evolving user expectations**:

  - *Problem*: User expectations and the context of feedback can evolve rapidly, making it challenging for static AI/ML models to remain accurate over time.

  - *Strategy*: Employ continuous learning approaches where AI/ML models are regularly updated with new data. This can involve techniques such as online learning, where models adapt in real time to changes in feedback trends and patterns.

- **Integration with existing systems**:

  - *Problem*: Seamlessly integrating AI/ML into existing feedback and development workflows can be technically challenging, potentially leading to disruptions in the feedback loop.

  - *Strategy*: Focus on AI/ML solutions that offer flexible integration capabilities with existing tools and platforms. Adopt a phased implementation approach to gradually introduce AI/ML functionalities, allowing for adjustment and optimization based on initial outcomes.

- **Ensuring user privacy and trust**:

  - *Problem*: Utilizing AI/ML to analyze user feedback raises concerns about user privacy and data security, potentially eroding user trust.

  - *Strategy*: Adopt and clearly communicate strict data privacy policies, ensuring that user feedback is analyzed in compliance with relevant regulations (e.g., GDPR). Employ privacy-preserving AI/ML techniques, such as federated learning or differential privacy, to analyze data without compromising individual privacy.

- **Overreliance on AI/ML insights**:

  - *Problem*: There might be an overreliance on AI/ML for decision-making, overlooking the importance of human intuition and understanding in interpreting feedback.

- *Strategy*: Establish guidelines that encourage a balanced approach to decision-making, combining AI/ML-generated insights with human judgment. Foster a culture that values the complementary roles of technology and human expertise in enhancing product and service quality.

- **Feedback volume and scalability**:

  - *Problem*: The sheer volume of feedback can overwhelm AI/ML systems, especially in scenarios where rapid scalability is required.

  - *Strategy*: Design AI/ML systems with scalability in mind, utilizing cloud-based solutions and distributed computing techniques to handle large datasets effectively. Regularly evaluate system performance and scalability, making necessary adjustments to infrastructure to meet demand.

By proactively addressing these challenges, organizations can effectively leverage AI/ML in their continuous feedback processes, ensuring that user and stakeholder feedback is accurately collected, analyzed, and acted upon to drive continuous improvement and innovation.

Real-world use case for AI/ML-assisted continuous feedback

A real-world example of using AI/ML-assisted tools for continuous feedback is illustrated by **Medallia**, a platform that leverages AI to analyze customer feedback across various channels in real time. Medallia's AI component, known as Medallia Athena, uses **natural language processing (NLP)** and ML to understand, categorize, and prioritize customer sentiments, opinions, and behaviors from sources such as surveys, social media, and direct customer interactions.

This technology enables businesses to automatically detect emerging trends, sentiment shifts, and potential issues in customer experiences as they happen, allowing companies to rapidly address concerns and capitalize on feedback. For example, if a negative trend in customer satisfaction begins to surface in certain regions or demographics, Medallia can alert managers immediately, enabling them to take swift action to resolve issues, enhance service quality, and improve customer satisfaction continuously. This real-time feedback processing and response mechanism is vital for businesses aiming to maintain a high level of customer engagement and satisfaction in dynamic markets.

Methodology for selecting AI/ML tools

Selecting the right AI/ML tools for continuous testing, quality, security, and feedback involves a comprehensive methodology that ensures the chosen tools align with organizational goals, integrate seamlessly with existing systems, and effectively address specific challenges within these domains. When distinguishing between generative AI tools (which generate new data or content) and predictive AI tools (which predict outcomes based on input data), the selection process must account for unique considerations related to the functionality, application, and potential impact of these technologies.

Figure 8.10 shows a structured methodology for selecting AI/ML tools, highlighting differences in selecting generative versus predictive AI tools.

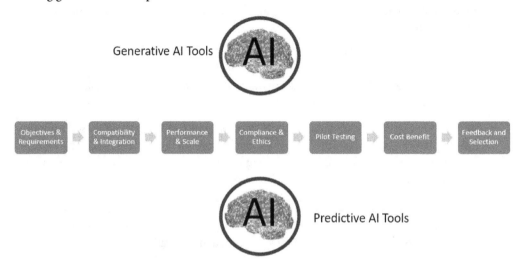

Figure 8.10 – Methodology for selecting AI/ML tools

Each of the steps is described in the following list:

1. **Define objectives and requirements**:

   - *For both*: Clearly outline what you aim to achieve by integrating AI/ML tools into continuous testing, quality, security, and feedback processes. Identify specific challenges and requirements, such as reducing false positives in security alerts or accelerating the feedback loop.

   - *Differences*: For generative AI tools, focus on creativity, content generation capabilities, and the tool's ability to produce diverse outputs. For predictive AI tools, prioritize accuracy, reliability, and the tool's capacity to handle vast datasets and provide actionable insights.

2. **Assess compatibility and integration**:

   - *For both*: Evaluate how well the AI/ML tools integrate with your existing development, testing, and deployment environments. Consider the compatibility with current workflows, data formats, and platforms.

   - *Differences*: Generative AI tools might require more robust creative input and output handling capabilities, while predictive AI tools often need strong data processing and analysis features that seamlessly integrate with your data sources and analytics platforms.

3. **Evaluate performance and scalability:**

    - *For both*: Test the tools' performance against benchmarks relevant to your objectives, including processing speed, accuracy, and scalability to handle growing data volumes and complexity.

    - *Differences*: For generative AI tools, assess the quality and relevance of generated content. For predictive AI tools, focus on the accuracy of predictions, the speed of data processing, and the model's performance under varying conditions.

4. **Review compliance and ethical considerations:**

    - *For both*: Ensure that the tools comply with data privacy, security regulations, and ethical guidelines. Consider the transparency of the AI/ML models and their decisions.

    - *Differences*: Generative AI tools may require additional scrutiny regarding the originality and copyright implications of generated content. Predictive AI tools might necessitate a deeper examination of potential biases in predictions and decision-making processes.

5. **Conduct pilot testing:**

    - *For both*: Implement a pilot project to test the selected AI/ML tools in a controlled environment. Monitor the impact on workflow efficiency, quality improvements, and user satisfaction.

    - *Differences*: For generative AI tools, pilot tests should focus on assessing the innovation, variety, and applicability of generated outputs. For predictive AI tools, the emphasis should be on the accuracy, timeliness, and relevance of predictions to real-world scenarios.

6. **Analyze cost benefits:**

    - *For both*: Evaluate the cost of implementation, training, and maintenance against the expected benefits, such as improved efficiency, enhanced security, or accelerated product development cycles.

    - *Differences*: Generative AI tools might involve costs related to creativity and content generation capabilities, which could yield new product features or content strategies. Predictive AI tools often require investment in data processing and analysis capabilities, which could lead to significant improvements in decision-making processes.

7. **Gather feedback and refine selection:**

    - *For both*: Collect feedback from pilot users and stakeholders to refine your tool selection. Consider ease of use, satisfaction with the results, and any unexpected challenges encountered.

    - *Differences*: Feedback for generative AI tools may center around the creativity and utility of generated content. For predictive AI tools, feedback might focus on the accuracy, usefulness, and actionable nature of predictions.

By following this methodology and acknowledging the nuances between selecting generative and predictive AI tools, organizations can make informed decisions that align with their strategic goals in continuous testing, quality, security, and feedback initiatives.

Summary

This chapter described AI and ML-enabled tools integrated within continuous testing, quality, security, and feedback practices. It explained how AI/ML technologies can greatly improve the efficiency, security, and responsiveness of development and delivery processes. The discussion offered a practical framework for automating and enhancing tasks with AI/ML-enabled tools.

Selecting the appropriate AI/ML tools is a critical step in effectively integrating these technologies. The differentiation between generative and predictive AI tools underscored the importance of a thoughtful approach in tool selection, ensuring that the chosen solutions align with organizational objectives and address the challenges of continuous processes. The chapter identified the challenges and considerations that accompany the application of AI/ML. Issues such as data quality, privacy concerns, and the need to maintain a balance between automation and human oversight are fundamental for the strategic implementation of AI/ML.

Looking forward, the next chapter will present use cases of continuous testing, quality, security, and feedback for organizations advancing their DevOps, DevSecOps, and SRE practices. Through practical examples and thorough analysis, it will demonstrate how AI/ML not only streamlines operations but also raises the standards of software development and maintenance, marking a significant shift toward more agile, resilient, and efficient digital transformation journeys.

Part 3: Deep Dive into Roadmaps, Implementation Patterns, and Measurements

Part 3 of this book shifts focus toward the practical aspects of applying continuous testing, quality, security, and feedback in the realms of DevOps, DevSecOps, and SRE. This section is structured to provide readers with a comprehensive understanding of how to bring these continuous strategies to life within their organizations.

Starting with real-world use cases, it showcases the transformative power of these practices across different operational contexts, offering insights into achieving higher levels of operational maturity. Following this, the book guides readers through the process of creating strategic roadmaps tailored to their organizational goals, ensuring a well-aligned digital transformation journey. It then explores various implementation patterns, offering structured approaches that have been proven to enhance the success of these strategic roadmaps. Finally, the section concludes by emphasizing the importance of measuring progress and outcomes, providing readers with the tools and frameworks necessary to track and evaluate the effectiveness of their continuous practices. This part of the book is essential for anyone looking to practically implement and benefit from continuous testing, quality, security, and feedback in their digital transformation efforts.

This part includes the following chapters:

- *Chapter 9, Use Cases for Integrating with DevOps, DevSecOps, and SRE*
- *Chapter 10, Building Roadmaps for Implementation*
- *Chapter 11, Understanding Transformation Implementation Patterns*
- *Chapter 12, Measuring Progress and Outcomes*

Use Cases for Integrating with DevOps, DevSecOps, and SRE

This chapter, as illustrated in *Figure 9.1*, describes practical applications of continuous testing, continuous quality, continuous security, and continuous feedback within the context of DevOps, DevSecOps, and SRE frameworks. The use cases illustrate how organizations can transform to higher levels of operational maturity.

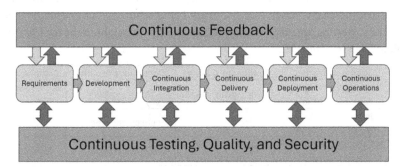

Figure 9.1 – Use cases for DevOps, DevSecOps, and SRE

This chapter is organized into several key sections. The first section, *Use cases for DevOps, DevSecOps, and SRE*, explains how the integration of these practices helps to create robust, resilient, and responsive IT ecosystems. The *Use cases with DevOps*, *Use cases with DevSecOps*, and *Use cases with SRE* sections explain how applications of each of the continuous methodologies work.

The chapter also explains how to effectively integrate continuous testing, quality, security, and feedback within DevOps, DevSecOps, and SRE practices.

The *Sustaining integrations* section offers a forward-looking perspective on maintaining and evolving these integrations over time, ensuring that organizations can adapt to emerging challenges while preserving the integrity and effectiveness of their operational practices.

By the end of this chapter, you will have not only explored a diverse array of use cases but also acquired the skills necessary to implement and sustain these integrations within your own organizations. Whether you're a seasoned practitioner or new to the world of DevOps, DevSecOps, and SRE, this chapter will equip you with the knowledge and insights needed to navigate the complexities of modern software delivery and operations, driving continuous improvement across every facet of your organization's technological landscape.

In this chapter, we'll cover the following main topics:

- Use cases for DevOps
- Use cases for DevSecOps
- Use cases for SRE
- Sustaining integrations

Let's get started!

Use cases for DevOps

The DevOps philosophy emphasizes the seamless integration of development and operations to enhance agility, speed, and quality of software delivery. A key to achieving this integration is embedding continuous practices – testing, quality, security, and feedback – throughout the life cycle of software development illustrated in *Figure 9.2*.

	Continuous Testing	Continuous Quality	Continuous Security	Continuous Feedback
Requirements Stage	Test requirements and coverage criteria	Establish quality criteria	Threat modeling	Stakeholders feedback on requirements
Development Stage	Unit Test and Test-Driven Development	Code scanners	Secure coding practices	Peer reviews
Continuous Integration	Automated integration level testing	Integration quality checks	Security scanning	Developer feedback
Continuous Delivery	Automated system-level tests	Quality release metrics	Security release metrics	Stakeholder release policies
Continuous Deploymenet	Deployment testing	Real-time quality monitoring	Real-time security monitoring	User feedback monitoring
Continuous Operations	Tests in production	Production quality monitoring	Monitor security patches	User feedback monitoring

Figure 9.2 – Use cases for DevOps

This section outlines how these continuous practices can be applied across the DevOps stages, ensuring a holistic approach to building and maintaining robust software systems.

Requirements stage

The *requirements stage* of a software value stream refers to the initial phase where the necessary specifications and expectations for a software product are defined and documented. During this stage, stakeholders, including business analysts, product managers, and customers, collaborate to articulate what the software should do, including functional requirements and non-functional requirements, such as performance, establishing clear and actionable requirements that guide the entire development process. This stage is crucial for setting the scope of the project, prioritizing features, and ensuring that the end product aligns with user needs and business goals.

- **Continuous testing use case – requirements validation**:

  - *Explanation*: Automated testing of requirements to validate their clarity, consistency, and testability before development begins.

  - *Importance*: Prevents misunderstandings or ambiguities in requirements that can lead to costly rework.

  - *Challenges*: Developing a framework that can test requirements in an automated fashion can be complex.

  - *Strategies*: Utilize **behavior-driven development (BDD)** frameworks to create testable specifications based on requirements.

- **Continuous quality use case – quality criteria definition**:

  - *Explanation*: Establishing clear quality criteria aligned with customer expectations and business goals.

  - *Importance*: Sets a benchmark for quality, guiding development, and testing efforts.

  - *Challenges*: Balancing comprehensive quality criteria with realistic project scopes and timelines.

  - *Strategies*: Collaborate with stakeholders to define essential quality attributes and incorporate feedback loops for ongoing refinement.

- **Continuous security use case – threat modeling**:

  - *Explanation*: Identifying potential security threats and vulnerabilities early in the lifecycle to guide the security posture of the project.

  - *Importance*: Ensures that security considerations are integrated from the start, reducing the risk of late-stage vulnerabilities.

  - *Challenges*: Conducting thorough threat modeling requires expertise and can be time-consuming; conducting threat modeling ahead of the design phase may not be sufficient for some designs and may need to wait until the design is completed.

- *Strategies*: Leverage automated threat modeling tools and conduct regular training to build security expertise within the team.

- **Continuous feedback use case – stakeholder feedback on requirements**:

  - *Explanation*: Gathering and incorporating feedback from stakeholders on the defined requirements and quality criteria.

  - *Importance*: Ensures that the project aligns with stakeholder expectations and business objectives.

  - *Challenges*: Managing and prioritizing feedback from diverse stakeholders

  - *Strategies*: Implement a structured feedback process and use prioritization matrices to address feedback effectively.

Development stage

The *development stage* of a software value stream is the phase where the actual coding and creation of the software take place. In this stage, developers translate the defined requirements into functional software through coding and initial unit testing. It is a critical phase where the foundation and core functionalities of the software are built, setting the stage for subsequent integration and testing activities that follow in the value stream.

- **Continuous testing use case – unit testing**:

  - *Explanation*: Developers write and execute unit tests alongside code development to validate individual functions or components.

  - *Importance*: Catches defects early, improving code quality and reducing downstream testing efforts.

  - *Challenges*: Ensuring comprehensive coverage and maintaining tests as code evolves.

  - *Strategies*: Adopt **test-driven development (TDD)** practices and leverage automated tools to track coverage and flag missing tests.

- **Continuous quality use case – code quality analysis**:

  - *Explanation*: Automated analysis of code quality using static code analysis tools to enforce coding standards and identify potential issues.

  - *Importance*: Maintains high code quality, readability, and maintainability.

  - *Challenges*: Configuring analysis tools to provide meaningful insights without overwhelming developers with false positives.

  - *Strategies*: Customize tool configurations based on project needs and review thresholds periodically based on feedback.

- **Continuous security use case – secure coding practices**:

    - *Explanation*: Integrating security best practices and guidelines into the development process to prevent common security flaws.

    - *Importance*: Reduces vulnerabilities in the software, enhancing the overall security posture.

    - *Challenges*: Keeping development teams up to date with evolving security practices and integrating security without impacting productivity.

    - *Strategies*: Provide regular security training and integrate security linting tools into the development environment.

- **Continuous feedback use case – peer code reviews**:

    - *Explanation*: Conducting regular peer code reviews to provide constructive feedback on code quality, design, and security.

    - *Importance*: Promotes knowledge sharing, improves code quality, and fosters a collaborative culture.

    - *Challenges*: Ensuring timely and effective reviews without slowing down the development process.

    - *Strategies*: Implement lightweight review processes and leverage tooling to automate routine checks.

Continuous integration stage

The *continuous integration stage* involves combining individual pieces of code developed by different team members into a single software system and conducting comprehensive testing to ensure that all components work together seamlessly. It's a critical checkpoint where issues such as conflicts between integrated components and bugs are identified and resolved, ensuring the software's functionality and stability before it moves on to the delivery phase, which prepares the product for release but does not include deployment.

- **Continuous testing use case – integration testing**:

    - *Explanation*: Automatically testing the integration of different components or systems to verify they work together as expected.

    - *Importance*: Identifies issues caused by the interaction between integrated components or systems.

    - *Challenges*: Designing and maintaining a comprehensive set of integration tests that cover all critical interactions.

    - *Strategies*: Use contract testing to validate interactions and mock external dependencies for more reliable and faster tests.

- **Continuous quality use case – build quality checks**:

  - *Explanation*: Integrating quality checks into the CI pipeline to automatically assess and enforce quality standards for each build.

  - *Importance*: Ensures that every build meets predefined quality standards before it progresses further in the pipeline.

  - *Challenges*: Defining and maintaining relevant and realistic quality metrics that accurately reflect the project's quality goals.

  - *Strategies*: Regularly review and adjust quality metrics based on project evolution and feedback from quality analysis tools.

- **Continuous security use case – automated security scanning**:

  - *Explanation*: Incorporating automated security scanning tools into the CI process to detect vulnerabilities and security issues in code and dependencies.

  - *Importance*: Identifies and mitigates security risks early in the development process.

  - *Challenges*: Integrating comprehensive security scans without significantly slowing down the CI pipeline.

  - *Strategies*: Implement incremental scanning and prioritize scans based on code changes and known risk areas.

- **Continuous feedback use case – build feedback loops**

  - *Explanation*: Providing immediate feedback to developers on the success or failure of integration attempts, including details on quality, security, and test results.

  - *Importance*: Enables developers to quickly address issues and learn from integration failures.

  - *Challenges*: Aggregating feedback from various tools and processes into actionable insights.

  - *Strategies*: Use dashboard and notification systems to consolidate feedback and prioritize issues based on severity and impact.

Continuous delivery stage

In the *continuous delivery stage* of a software value stream, every change that passes the automated tests in the integration stage is automatically prepared for a release to production. This involves packaging the software, conducting further automated tests to validate the release, and ensuring that the software is always deployable. This stage is crucial for maintaining a steady flow of updates ready for deployment, enabling quick and efficient delivery of new features and fixes to users.

- **Continuous testing use case – pre-release testing**:

  - *Explanation*: Executing a comprehensive suite of automated tests in a staging environment that mirrors production to validate the release.

  - *Importance*: Ensures that the release candidate is thoroughly tested in an environment that closely resembles production.

  - *Challenges*: Maintaining an up-to-date and accurate staging environment that includes all necessary data and configurations.

  - *Strategies*: Use **Infrastructure as Code (IaC)** to manage and replicate production-like environments and implement blue-green deployments to minimize differences.

- **Continuous quality use case – release quality gates**:

  - *Explanation*: Defining and enforcing quality gates that a release must pass before it can be deployed to production, based on test results, quality metrics, and security scans.

  - *Importance*: Guarantees that only releases meeting all quality and security criteria are deployed, reducing the risk of production issues.

  - *Challenges*: Balancing strict quality criteria with the need for rapid delivery and avoiding bottlenecks in the release process.

  - *Strategies*: Continuously evaluate and adjust quality gates to align with evolving project needs and risk tolerance.

- **Continuous security use case – security review and approval**:

  - *Explanation*: Conducting a final security review and obtaining approval from the security team before a release is deployed to production.

  - *Importance*: Ensures that all security concerns have been addressed and that the release complies with organizational security policies.

  - *Challenges*: Coordinating between development, security, and operations teams to ensure timely reviews without delaying releases.

  - *Strategies*: Integrate automated security checks and risk assessments into the delivery process to streamline security reviews.

- **Continuous feedback use case – stakeholder review and sign-off**:

  - *Explanation*: Gathering final approval from stakeholders based on the release's compliance with requirements, quality standards, and security policies.

  - *Importance*: Ensures that the release aligns with business goals and stakeholder expectations, minimizing the risk of post-deployment issues.

- *Challenges*: Managing expectations and communications with a diverse group of stakeholders to obtain timely sign-off.

- *Strategies*: Use automated release notes and dashboards to provide transparent updates on release status and facilitate easier stakeholder reviews.

Continuous deployment stage

The *continuous deployment stage* of a software value stream involves automatically deploying every change that has successfully passed through the continuous delivery process to production, ensuring that updates are frequently and predictably released. During this phase, deployment testing is crucial to verify that the release functions correctly in the live environment. If these tests fail, the system is designed to automatically roll back to the previous stable version to maintain service continuity and minimize disruptions. This automated process enhances responsiveness to market and user demands by ensuring a constant flow of improvements and fixes.

- **Continuous testing use case – canary testing**:

  - *Explanation*: Gradually rolling out changes to a small subset of users to monitor performance and detect issues before a full rollout.

  - *Importance*: Allows for the detection and mitigation of potential issues in a controlled manner, reducing the impact on the broader user base.

  - *Challenges*: Designing effective canary tests that provide meaningful feedback without affecting the user experience.

  - *Strategies*: Utilize feature flags and automated monitoring tools to manage and analyze canary deployments.

- **Continuous quality use case – continuous monitoring for quality metrics**:

  - *Explanation*: Implementing real-time monitoring of quality metrics in production to ensure ongoing compliance with quality standards.

  - *Importance*: Provides continuous insight into the quality of the application in production, allowing for rapid response to any degradation.

  - *Challenges*: Selecting meaningful quality metrics and setting appropriate thresholds for alerts.

  - *Strategies*: Regularly review and adjust quality metrics and thresholds based on historical performance data and feedback.

- **Continuous security use case – continuous compliance monitoring**:

  - *Explanation*: Monitoring compliance with security policies and regulations in real time to ensure continuous adherence.

- *Importance*: Maintains the application's compliance status and reduces the risk of security breaches and regulatory issues.

- *Challenges*: Implementing comprehensive compliance monitoring that covers all relevant regulations and policies without overwhelming the team with false positives.

- *Strategies*: Use specialized compliance monitoring tools and regularly update compliance rules to reflect changes in regulations.

- **Continuous feedback use case – user behavior analysis**:

  - *Explanation*: Collecting and analyzing user behavior data to gather feedback on the deployed features and identify areas for improvement.

  - *Importance*: Offers direct insight into how users interact with the application, highlighting potential areas for enhancement or adjustment.

  - *Challenges*: Ensuring user privacy and data security while collecting and analyzing behavior data.

  - *Strategies*: Implement clear data collection policies, use anonymization techniques, and obtain user consent where necessary.

Continuous operations stage

The *continuous operations stage* continues until the next version of the software is deployed. During this phase, the software is under continuous monitoring and maintenance to ensure it performs optimally. Practices such as continuous testing, quality assurance, security monitoring, and gathering feedback are integral to this stage. They help promptly identify and resolve any operational issues, maintain security standards, and incorporate user feedback into future updates. This continuous cycle supports the software in delivering consistent performance and security while adapting to evolving user requirements until the next deployment.

- **Continuous testing use case – post-deployment testing**:

  - *Explanation*: Conducting automated tests in the production environment after deployment to verify that the application performs as expected under real-world conditions.

  - *Importance*: Confirms the success of the deployment and the operational stability of the application.

  - *Challenges*: Implementing non-disruptive tests that can run in production without affecting user experience.

  - *Strategies*: Use synthetic monitoring and shadow traffic to simulate real user interactions without impacting actual users.

- **Continuous quality use case – ongoing quality improvement**:

  - *Explanation*: Leveraging feedback and data collected from production to continuously improve the quality of the application.

  - *Importance*: Ensures that the application evolves in response to user needs and feedback, maintaining high standards of quality.

  - *Challenges*: Prioritizing quality improvement initiatives among ongoing development and operational tasks.

  - *Strategies*: Implement a structured process for reviewing feedback, conducting root cause analyses, and integrating quality improvements into the development backlog.

- **Continuous security use case – continuous security patching**:

  - *Explanation*: Implementing an automated process for applying security patches and updates to the application and its dependencies.

  - *Importance*: Keeps the application secure against known vulnerabilities, reducing the risk of exploitation.

  - *Challenges*: Ensuring that patches do not introduce new issues or degrade application performance.

  - *Strategies*: Use automated testing and canary deployments for patches to validate their safety before wider rollout.

- **Continuous feedback use case – real-time user feedback collection**:

  - *Explanation*: Utilizing tools and processes to collect and analyze user feedback in real time, allowing for immediate action on issues and suggestions.

  - *Importance*: Enables the organization to rapidly respond to user needs and enhance the user experience.

  - *Challenges*: Managing and prioritizing a large volume of real-time feedback from diverse sources.

  - *Strategies*: Implement feedback management systems to aggregate and prioritize feedback and integrate feedback analysis into regular development cycles.

Integrating continuous testing, quality, security, and feedback into each stage of the DevOps life cycle is essential for developing and maintaining high-quality, secure, and user-centered software. While challenges exist in implementing these continuous practices, the strategies outlined provide a framework for overcoming these hurdles, ensuring that the benefits of a DevOps approach are fully realized. By adopting these practices, organizations can enhance their agility, improve software quality, bolster security measures, and respond more effectively to user needs, driving success in today's competitive and fast-paced digital landscape.

Real-world use case for DevOps

A real-world use case that effectively integrates continuous testing, quality, security, and feedback practices into a DevOps pipeline can be seen in a financial services company aiming to enhance its software deployment frequency and reliability. In this scenario, the company utilizes Jenkins as a core part of its DevOps pipeline to automate and manage continuous integration and delivery processes. Jenkins orchestrates a series of stages including building code, running automated tests, and deploying to staging environments.

Continuous testing is implemented using Selenium for automated regression and functional testing to ensure each new release functions as intended without breaking existing features. For quality assurance, SonarQube is integrated to perform static code analysis, identifying potential bugs and vulnerabilities to maintain high code quality standards.

In terms of security, the OWASP ZAP tool is used to conduct automated security scans during the CI/CD process, ensuring that vulnerabilities are identified and addressed before deployment. Additionally, feedback mechanisms are enhanced by integrating feature flagging tools such as LaunchDarkly, which allow real-time user feedback on new features in production without impacting all users. This integration not only speeds up the feedback loop but also allows for safer, gradual rollouts and quick adaptation based on user input.

This integration of tools and practices ensures that the company's software releases are more secure, reliable, and aligned with user expectations, thereby significantly enhancing the overall software delivery life cycle.

Use cases for DevSecOps

The adoption of DevSecOps is pivotal in today's software development and deployment practices, emphasizing the seamless integration of security measures throughout the DevOps life cycle, as illustrated in *Figure 9.3*.

	Continuous Testing	Continuous Quality	Continuous Security	Continuous Feedback
Requirements Stage	Security requirements and coverage criteria	Security standards	Threat modeling	Stakeholders alignment on security priorities
Development Stage	Secure coding practices	Security peer reviews	Security unit tests	Developer feedback on security practices
Continuous Integration	Security scanning	Integration security metrics checks	Vulnerabilities scanning	Developer securiy feedback
Continuous Delivery	Automated system-level security tests	Security release metrics	Security release metrics	Stakeholder security release policies
Continuous Deploymenet	Deployment security testing	Real-time security monitoring	Real-time security monitoring	User security feedback monitoring
Continuous Operations	Security tests in production	Production security monitoring	Monitor security patches	User security feedback monitoring

Figure 9.3 – Use cases for DevSecOps

This approach not only accelerates delivery times but also ensures that security is a foundational component rather than an afterthought. This section delves into the application of continuous testing, continuous quality, continuous security, and continuous feedback across each stage of the DevSecOps life cycle: requirements, development, continuous integration, continuous delivery, continuous deployment, and continuous operations. For each stage, we explore distinct use cases, their significance, associated challenges, and strategies to address these challenges effectively.

Requirements stage

- **Continuous testing use case – security requirements validation:**

    - *Explanation*: Automating the validation of security requirements ensures that they are testable and aligned with security policies and standards.

    - *Importance*: Sets a clear security baseline for the project, guiding subsequent development and testing efforts.

    - *Challenges*: Developing automated tools or scripts that can effectively validate nuanced security requirements.

    - *Strategies*: Leverage BDD techniques to define security requirements in a testable format.

- **Continuous quality use case – quality and security standards definition:**

    - *Explanation*: Establishing comprehensive quality and security standards that the project must adhere to throughout its life cycle.

    - *Importance*: Provides a clear set of criteria for evaluating the quality and security of the software, ensuring consistency and compliance.

    - *Challenges*: Balancing stringent quality and security standards with practical development and deployment timelines.

    - *Strategies*: Develop incremental standards that can evolve as the project progresses, allowing for flexibility without compromising on core principles.

- **Continuous security use case – threat modeling and risk assessment:**

    - *Explanation*: Conducting threat modeling and risk assessments to identify potential security threats and vulnerabilities early in the life cycle.

    - *Importance*: Informs the design and development process by highlighting areas that require additional security measures.

    - *Challenges*: Requires deep security expertise and understanding of potential attack vectors relevant to the project.

    - *Strategies*: Utilize automated threat modeling tools and incorporate external security experts for comprehensive assessments.

- **Continuous feedback use case – stakeholder security expectations alignment**:

  - *Explanation*: Engaging with stakeholders to align expectations regarding security postures and priorities.

  - *Importance*: Ensures that the project's security measures meet or exceed stakeholder expectations, avoiding late-stage misunderstandings.

  - *Challenges*: Facilitating effective communication between technical teams and stakeholders with varying levels of security knowledge.

  - *Strategies*: Use clear, non-technical language to describe security measures and their implications for project outcomes.

Development stage

- **Continuous testing use case – secure coding practices**:

  - *Explanation*: Implementing secure coding practices involves leveraging static and dynamic analysis tools to detect vulnerabilities in code as it is written.

  - *Importance*: Reduces the number of vulnerabilities in the code base, enhancing the overall security posture of the application.

  - *Challenges*: Integrating these tools into the development workflow without disrupting developer productivity.

  - *Strategies*: Integrate security tools into the IDE and provide developers with immediate feedback to correct issues as they code.

- **Continuous quality use case – code review and analysis for quality assurance**:

  - *Explanation*: Conducting automated and manual code reviews to ensure adherence to quality and security standards.

  - *Importance*: Promotes high code quality and security by facilitating peer oversight and leveraging automated tools to detect issues.

  - *Challenges*: Managing the additional time and resources required for thorough code reviews.

  - *Strategies*: Automate routine checks with static analysis tools and prioritize manual review efforts for complex or critical parts of the code base.

- **Continuous security use case – security unit testing**:

  - *Explanation*: Developing and executing unit tests specifically designed to validate security functions and controls within the application.

  - *Importance*: Ensures that security mechanisms are functioning as intended, protecting against known vulnerabilities and attack vectors.

- *Challenges*: Creating comprehensive test cases that cover a wide range of security scenarios.
- *Strategies*: Use a combination of manual security expertise and automated tools to generate and execute security-focused unit tests.

- **Continuous feedback use case – developer feedback on security practices**:

  - *Explanation*: Collecting and acting on feedback from developers regarding the effectiveness and efficiency of implemented security practices.
  - *Importance*: Improves the security posture of the application by refining practices based on real-world development experiences.
  - *Challenges*: Encouraging open and constructive feedback from developers who may not be security experts.
  - *Strategies*: Establish a culture of open communication and continuous learning, emphasizing the value of security within the development process.

Continuous integration stage

- **Continuous testing use case – automated security scanning**:

  - *Explanation*: Integrating automated security scanning tools into the CI pipeline to detect vulnerabilities and misconfigurations as new code is integrated.
  - *Importance*: Identifies and addresses security issues early in the development process, reducing the risk of vulnerabilities making it to production.
  - *Challenges*: Balancing the thoroughness of scans with the need to maintain a rapid CI pipeline.
  - *Strategies*: Implement incremental scanning and prioritize scanning based on changes to the code base and known high-risk areas.

- **Continuous quality use case – build quality assurance**:

  - *Explanation*: Ensuring each build meets predefined quality criteria by integrating quality checks into the CI process.
  - *Importance*: Maintains a consistent level of quality throughout the development process, preventing quality degradation over time.
  - *Challenges*: Defining and enforcing quality criteria that are both comprehensive and realistic.
  - *Strategies*: Use automated tools to measure and report on quality metrics, and adjust criteria based on historical data and project evolution.

- **Continuous security use case – dependency scanning**:

  - *Explanation*: Automatically scanning dependencies for known vulnerabilities as part of the CI process.

  - *Importance*: Prevents the introduction of vulnerabilities through third-party libraries and frameworks, a common attack vector.

  - *Challenges*: Keeping up with the latest vulnerability databases and managing false positives.

  - *Strategies*: Integrate real-time vulnerability feeds into scanning tools and establish a process for reviewing and acting on scan results.

- **Continuous feedback use case – CI pipeline feedback for security and quality**:

  - *Explanation*: Providing immediate feedback to developers on security and quality issues identified during the CI process.

  - *Importance*: Enables rapid remediation of issues, fostering a culture of continuous improvement.

  - *Challenges*: Aggregating feedback from multiple tools into actionable insights for developers.

  - *Strategies*: Use dashboards and integrated development tools to consolidate and prioritize feedback, highlighting critical issues for immediate action.

Continuous delivery stage

- **Continuous testing use case – security and quality gateways**:

  - *Explanation*: Implementing security and quality gateways as part of the CD process to ensure that only code meeting strict criteria is promoted to later stages.

  - *Importance*: Acts as a final check to prevent insecure or low-quality code from being deployed to production.

  - *Challenges*: Defining appropriate thresholds for passing gateways without impeding the delivery process.

  - *Strategies*: Establish clear, measurable criteria for gateways and regularly review them based on project outcomes and stakeholder feedback.

- **Continuous quality use case – release candidate testing**:

  - *Explanation*: Conducting comprehensive security testing on release candidates to validate both functionality and adherence to security standards.

  - *Importance*: Ensures that releases are both functionally complete and meet the project's quality benchmarks.

- *Challenges*: Efficiently executing a broad suite of tests without delaying the release process.

- *Strategies*: Automate as much of the testing process as possible and prioritize testing based on risk and impact assessments.

- **Continuous security use case – pre-deployment security review**:

  - *Explanation*: Performing a final security review and approval process before deployment to validate compliance with security policies.

  - *Importance*: Ensures that the release adheres to all organizational and regulatory security requirements.

  - *Challenges*: Conducting thorough reviews in a timely manner to avoid delaying deployments.

  - *Strategies*: Streamline the review process through automation and predefined checklists, and ensure that security personnel are integrated into the DevSecOps team.

- **Continuous feedback use case – release feedback collection**:

  - *Explanation*: Gathering feedback from stakeholders on the release process, including security and quality aspects, to inform future improvements.

  - *Importance*: Identifies areas for process refinement and enhancement, driving continuous improvement in delivery practices.

  - *Challenges*: Effectively collating and acting on feedback from a diverse range of stakeholders.

  - *Strategies*: Implement structured feedback mechanisms and regular review meetings to discuss feedback and plan actions.

Continuous deployment stage

- **Continuous testing use case – post-deployment verification**:

  - *Explanation*: Automatically verifying the integrity and security of the deployment once it is live in the production environment.

  - *Importance*: Confirms the successful deployment of the application and its operational security posture.

  - *Challenges*: Performing comprehensive verification without impacting the live environment or user experience.

  - *Strategies*: Utilize canary deployments and synthetic transactions to validate deployments in a controlled manner.

- **Continuous quality use case – production environment monitoring**:

  - *Explanation*: Continuously monitoring the production environment to ensure that it meets quality and performance standards.

  - *Importance*: Maintains high levels of user satisfaction and system reliability by proactively identifying and addressing quality issues.

  - *Challenges*: Filtering through vast amounts of monitoring data to identify actionable insights.

  - *Strategies*: Implement intelligent monitoring solutions that leverage machine learning to prioritize issues based on severity and impact.

- **Continuous security use case – real-time threat detection**:

  - *Explanation*: Employing real-time monitoring and threat detection tools to identify and respond to security incidents in the production environment.

  - *Importance*: Minimizes the potential impact of security breaches by enabling rapid detection and response.

  - *Challenges*: Balancing sensitivity and specificity to minimize false positives while ensuring that no significant threats are overlooked.

  - *Strategies*: Fine-tune detection algorithms based on historical incident data and regularly update threat intelligence sources.

- **Continuous feedback use case – user experience monitoring**:

  - *Explanation*: Collecting and analyzing user feedback and usage data to understand the real-world experience of the application.

  - *Importance*: Provides insights into how users interact with the application and where improvements can be made to enhance satisfaction.

  - *Challenges*: Integrating feedback from multiple channels and translating it into actionable improvements.

  - *Strategies*: Implement comprehensive user analytics and feedback tools, and establish cross-functional teams to address feedback in an agile manner.

Continuous operations stage

- **Continuous testing use case – operational readiness testing**:

  - *Explanation*: Conducting ongoing testing to ensure that the system remains ready to handle operational demands, including load testing and disaster recovery simulations.

  - *Importance*: Guarantees the system's resilience and reliability under varying operational conditions.

- *Challenges*: Simulating realistic operational scenarios without impacting production systems or user experience.

- *Strategies*: Utilize shadow traffic and simulation tools to mimic real user behaviors and operational conditions in a controlled environment.

- **Continuous quality use case – quality improvement initiatives**:

  - *Explanation*: Implementing continuous quality improvement initiatives based on feedback and monitoring data to enhance system performance and user satisfaction.

  - *Importance*: Ensures that the application evolves to meet user needs and quality expectations over time.

  - *Challenges*: Prioritizing and implementing improvements in a way that balances new feature development with quality enhancements.

  - *Strategies*: Adopt a data-driven approach to identify quality improvement opportunities and integrate them into the product roadmap.

- **Continuous security use case – continuous compliance and security posture management**:

  - *Explanation*: Continuously monitoring and adjusting the security posture of the application to ensure compliance with evolving security policies and standards.

  - *Importance*: Maintains the application's security integrity over time, adapting to new threats and compliance requirements.

  - *Challenges*: Keeping abreast of changes in security standards and regulatory requirements and implementing changes without disrupting operations.

  - *Strategies*: Leverage automated compliance monitoring tools and establish a dedicated team to oversee security posture management.

- **Continuous feedback use case – operational feedback loop**:

  - *Explanation*: Establishing a feedback loop with operations teams to continuously improve operational practices based on real-world experiences and incidents.

  - *Importance*: Enhances operational efficiency and reliability by incorporating lessons learned from the front lines.

  - *Challenges*: Effectively capturing and integrating feedback from operations into continuous improvement processes.

  - *Strategies*: Implement regular retrospectives and debriefs following incidents, and integrate operational feedback into development and security practices.

Incorporating continuous testing, quality, security, and feedback across the DevSecOps life cycle represents a comprehensive approach to software development and deployment, ensuring that security is integrated at every stage. While challenges in implementation exist, strategic approaches (ranging from automation and tool integration to stakeholder engagement and continuous learning) can effectively address these obstacles. By embracing these continuous practices, organizations can not only enhance their security posture but also improve overall software quality and operational efficiency, driving forward in the competitive landscape of software development with confidence and resilience.

Real-world use case for DevSecOps

A real-world use case involving the integration of continuous testing, quality, security, and feedback practices into a DevSecOps pipeline can be seen within a healthcare software provider aiming to ensure the security and efficiency of its patient data management system. In this scenario, the company leverages GitHub Actions to automate workflows, which helps streamline both development and operations processes within a single platform.

Continuous testing is carried out with Cypress, an end-to-end testing framework, to automatically test every code commit and pull request for functional and integration issues. For quality, the company uses ESLint integrated into its development environment to maintain high code standards and catch syntax errors and problematic patterns in its JavaScript code.

Security is a critical concern due to the sensitive nature of health data. The company incorporates Aqua Security into the pipeline to scan container images for vulnerabilities and enforce security policies across the development life cycle. Feedback is gathered continuously through tools such as Jira, where bugs and user feedback from the production environment are logged and linked directly to development tasks. This facilitates rapid responses to issues and helps prioritize development work based on user impact.

By integrating these tools and practices into their DevSecOps pipeline, the healthcare software provider enhances its ability to deliver secure, high-quality software faster, while staying compliant with stringent industry regulations.

Use cases for SRE

Site reliability engineering (**SRE**) embodies a discipline that applies aspects of software engineering to solve problems in infrastructure and operations, focusing on the reliability, scalability, and maintainability of systems. SRE integrates continuous testing, continuous quality, continuous security, and continuous feedback into the life cycle of system development and operations, as illustrated in *Figure 9.4*.

	Continuous Testing	Continuous Quality	Continuous Security	Continuous Feedback
Requirements Stage	Reliability and testability requirements	Quality requirements	Security requirements	Feedback requirements
Development Stage	Unit and TDD for reliability cases	Quality metrics	Security unit tests	Developer feedback
Continuous Integration	Automated reliability tests in Continuous Integration	Integration quality metrics checks	Security scanning	Developer feedback on integration results
Continuous Delivery	Automated system level security tests	Release quality metrics	Release security metrics	Stakeholder release policies
Continuous Deploymenet	Deployment testing	Real-time quality feedback during deploy	Real-time security feedback during deploy	User satisfaction feedback during deploy
Continuous Operations	Traffic and capacity testing in production	Real-time quality feedback in production	Monitor security patches in production	User feedback monitoring in production

Figure 9.4 – Use cases for SRE

This section explores the implementation of these continuous practices across each stage of the SRE life cycle: requirements, development, continuous integration, continuous delivery, continuous deployment, and continuous operations. Each stage presents unique use cases, highlighting their significance, the challenges involved, and strategies for overcoming these challenges.

Requirements stage

- **Continuous testing use case – reliability requirements validation**:

  - *Explanation*: Ensuring that reliability requirements are clearly defined, measurable, and testable through automated testing.

  - *Importance*: Sets a concrete foundation for reliability engineering throughout the system's life cycle.

  - *Challenges*: Quantifying and creating automated tests for reliability requirements can be complex.

  - *Strategies*: Utilize **service-level objectives** (**SLOs**) and error budgets as measurable indicators of reliability and incorporate them into automated testing frameworks.

- **Continuous quality use case – quality attributes specification**:

  - *Explanation*: Defining specific quality attributes related to reliability, maintainability, and scalability.

  - *Importance*: Guides the design and development process toward building a high-quality system that meets user and business needs.

  - *Challenges*: Balancing various quality attributes without compromising on functionality.

  - *Strategies*: Prioritize quality attributes based on stakeholder input and potential impact on system reliability and performance.

- **Continuous security use case – security requirements planning**:

  - *Explanation*: Identifying and integrating security requirements and controls based on the system's threat model.

  - *Importance*: Ensures that security considerations are baked into the system from the start, minimizing vulnerabilities.

  - *Challenges*: Keeping up with evolving security threats and ensuring comprehensive coverage.

  - *Strategies*: Conduct regular threat modeling sessions and update security requirements based on emerging threats and vulnerabilities.

- **Continuous feedback use case – requirements feedback loop**:

  - *Explanation*: Establishing mechanisms for collecting and incorporating feedback on requirements from stakeholders and end users.

  - *Importance*: Ensures that the system's requirements remain aligned with user needs and business objectives.

  - *Challenges*: Efficiently managing and prioritizing a large volume of feedback.

  - *Strategies*: Implement structured feedback collection processes and regular review cycles to update requirements based on feedback.

Development stage

- **Continuous testing use case – unit and integration testing for reliability**:

  - *Explanation*: Developing unit and integration tests that focus on reliability aspects of the system, such as fault tolerance and graceful degradation.

  - *Importance*: Early identification and resolution of reliability issues during the development phase.

  - *Challenges*: Creating realistic test scenarios that accurately represent production environments.

  - *Strategies*: Leverage chaos engineering principles to simulate real-world conditions and integrate resilience testing into the development process.

- **Continuous quality use case – code quality standards**:

  - *Explanation*: Enforcing code quality standards through automated linting, code reviews, and static analysis tools.

  - *Importance*: Maintains high code quality, enhancing the system's maintainability and reliability.

  - *Challenges*: Ensuring adherence to quality standards without hindering development speed.

  - *Strategies*: Automate quality checks where possible and integrate them into the developers' workflow to provide immediate feedback.

- **Continuous security use case – secure development practices**:

  - *Explanation*: Incorporating secure coding practices and tools into the development process to identify and remediate security issues early.

  - *Importance*: Reduces the introduction of security vulnerabilities during development.

  - *Challenges*: Keeping development teams updated on the latest security practices and integrating them seamlessly into the development process.

  - *Strategies*: Provide ongoing security training and integrate security tools directly into the development environment.

- **Continuous feedback use case – developer experience feedback**:

  - *Explanation*: Gathering feedback from the development team on tools, practices, and processes to identify areas for improvement.

  - *Importance*: Enhances the development process and tooling, leading to increased productivity and job satisfaction.

  - *Challenges*: Creating a culture where feedback is regularly given and acted upon.

  - *Strategies*: Implement regular retrospectives and anonymous feedback tools to encourage open communication.

Continuous integration stage

- **Continuous testing use case – automated reliability testing in CI**:

  - *Explanation*: Integrating automated reliability testing into the CI pipeline, including load testing and chaos experiments.

  - *Importance*: Ensures that every change is vetted for its impact on the system's reliability before merging.

  - *Challenges*: Designing and maintaining an effective suite of reliability tests that can run quickly and provide meaningful feedback.

  - *Strategies*: Focus on incremental testing and prioritize tests based on historical issues and high-risk areas.

- **Continuous quality use case – continuous code quality monitoring**:

  - *Explanation*: Utilizing tools within the CI pipeline to continuously monitor and report on code quality metrics.

  - *Importance*: Keeps code quality consistently high, facilitating easier maintenance and scaling.

- *Challenges*: Defining meaningful quality metrics that accurately reflect the code base's health.

- *Strategies*: Use a combination of static code analysis, code complexity metrics, and code coverage reports to monitor quality.

- **Continuous security use case – CI security scanning**:

  - *Explanation*: Performing security scanning and analysis as part of the CI process to identify vulnerabilities early.

  - *Importance*: Minimizes the risk of deploying insecure code to later stages or production.

  - *Challenges*: Integrating comprehensive security scanning without significantly slowing down the CI process.

  - *Strategies*: Use incremental and differential scanning techniques to focus on newly introduced changes and known high-risk areas.

- **Continuous feedback use case – CI feedback for improvement**:

  - *Explanation*: Providing immediate feedback to developers on the results of CI processes, including test failures, security scan results, and quality metrics.

  - *Importance*: Enables developers to quickly address issues, improving the code base's overall health and security posture.

  - *Challenges*: Aggregating feedback from multiple sources into actionable insights.

  - *Strategies*: Use integrated dashboards and notification systems to consolidate feedback and prioritize based on severity and impact.

Continuous delivery stage

- **Continuous testing use case – pre-deployment testing**:

  - *Explanation*: Conducting comprehensive testing in a staging environment that mirrors production to validate changes before deployment.

  - *Importance*: Identifies any issues in a controlled environment, reducing the risk of production incidents.

  - *Challenges*: Ensuring the staging environment accurately reflects production in terms of data, scale, and configuration.

  - *Strategies*: Use IaC to manage environment configurations and keep staging and production environments in sync.

- **Continuous quality use case – release quality assurance**:

  - *Explanation*: Verifying that each release meets established quality benchmarks before deployment.

  - *Importance*: Ensures that only high-quality changes are deployed, maintaining the integrity and reliability of the production environment.

  - *Challenges*: Defining and enforcing quality benchmarks that are realistic yet stringent enough to uphold high standards.

  - *Strategies*: Automate quality checks and gate deployments based on quality metrics, incorporating manual review processes for critical releases.

- **Continuous security use case – security review and compliance checks**:

  - *Explanation*: Performing a final security review and compliance check before deployment to ensure that the release meets all security requirements and regulations.

  - *Importance*: Prevents security breaches and compliance violations, protecting the organization and its users.

  - *Challenges*: Conducting thorough reviews quickly enough to not delay deployments.

  - *Strategies*: Automate as much of the compliance and security review process as possible, using policy as code to enforce security standards.

- **Continuous feedback use case – pre-deployment stakeholder feedback**:

  - *Explanation*: Collecting feedback from stakeholders on upcoming changes and releases to ensure that they align with business goals and user expectations.

  - *Importance*: Aligns development efforts with business objectives and user needs, increasing the value of delivered changes.

  - *Challenges*: Efficiently gathering and incorporating feedback from a broad group of stakeholders.

  - *Strategies*: Use feature flagging and canary releases to gather feedback on changes in a controlled manner before full deployment.

Continuous deployment stage

- **Continuous testing use case – automated deployment testing**:

  - *Explanation*: Automatically testing the deployment process and the application post-deployment to validate successful deployment and operational status.

  - *Importance*: Ensures that the deployment process is reliable and that the application remains functional after deployment.

- *Challenges*: Creating tests that can accurately verify the success of deployments and the operational status of the application in production.

- *Strategies*: Use smoke testing and synthetic transactions to validate critical functionalities post-deployment.

- **Continuous quality use case – continuous monitoring for quality metrics**:

  - *Explanation*: Continuously monitoring application and infrastructure quality metrics in production to ensure that they meet predefined standards.

  - *Importance*: Provides real-time insight into the impact of deployments on application quality and user experience.

  - *Challenges*: Determining which quality metrics are most indicative of user experience and system performance.

  - *Strategies*: Implement comprehensive monitoring solutions that track a wide range of metrics, using machine learning to identify anomalies and trends.

- **Continuous security use case – continuous security monitoring**:

  - *Explanation*: Implementing continuous monitoring for security threats and vulnerabilities in the production environment.

  - *Importance*: Enables real-time detection and response to security incidents, minimizing potential damage.

  - *Challenges*: Filtering and prioritizing alerts to manage the volume of security data effectively without missing critical threats.

  - *Strategies*: Utilize advanced **security information and event management (SIEM)** systems and integrate automated response mechanisms.

- **Continuous feedback use case – real-time user feedback collection**:

  - *Explanation*: Collecting and analyzing user feedback in real time to assess the impact of deployments on user satisfaction and experience.

  - *Importance*: Provides immediate insights into how changes affect users, allowing for quick adjustments if necessary.

  - *Challenges*: Integrating feedback collection mechanisms into the user experience without being intrusive.

  - *Strategies*: Implement unobtrusive feedback tools, such as in-app surveys and usage analytics, to gather insights directly from users.

Continuous operations stage

- **Continuous testing use case – production traffic and load testing**:

  - *Explanation*: Simulating production traffic and load to test the system's capacity and scalability under real-world conditions.

  - *Importance*: Ensures that the system can handle expected and peak loads, preventing downtime and degraded performance.

  - *Challenges*: Simulating realistic traffic patterns and volumes without impacting actual users.

  - *Strategies*: Use shadow traffic and load testing tools that replicate real user behaviors and traffic patterns.

- **Continuous quality use case – ongoing quality improvement**:

  - *Explanation*: Leveraging continuous feedback and monitoring data to drive ongoing quality improvements in the system.

  - *Importance*: Ensures that the system continually evolves to meet user needs and quality standards, improving satisfaction and reliability.

  - *Challenges*: Prioritizing quality improvements among new feature development and operational tasks.

  - *Strategies*: Adopt a structured process for quality improvement, incorporating feedback and monitoring insights into the product roadmap.

- **Continuous security use case – continuous vulnerability management**:

  - *Explanation*: Implementing an ongoing process for identifying, assessing, and mitigating vulnerabilities in the production environment.

  - *Importance*: Protects against evolving security threats by ensuring that vulnerabilities are promptly addressed.

  - *Challenges*: Keeping up with new vulnerabilities and efficiently managing the remediation process.

  - *Strategies*: Utilize automated vulnerability scanning tools and integrate patch management into the operational workflow.

- **Continuous feedback use case – operational feedback and incident analysis**:

  - *Explanation*: Analyzing operational feedback and incidents to identify root causes and implement preventative measures.

  - *Importance*: Improves system reliability and operational efficiency by learning from operational experiences and incidents.

- *Challenges*: Systematically capturing and analyzing feedback and incidents to derive remediation and future mitigation actions.

- *Strategies*: Utilize blameless post-mortems to determine future mitigations.

The integration of continuous testing, continuous quality, continuous security, and continuous feedback into the SRE life cycle embodies a holistic approach to building and maintaining reliable, secure, and high-quality systems. While each stage presents distinct challenges, the strategic application of these continuous practices enables organizations to enhance their operational efficiency, security posture, and user satisfaction. By fostering a culture of continuous improvement and learning, SRE teams can drive technological and operational excellence, ensuring that their systems are resilient, scalable, and aligned with user needs and business goals.

Real-world use case for SRE

A real-world use case that integrates continuous testing, quality, security, and feedback practices into SRE practices can be seen at an e-commerce company aiming to optimize the reliability and user experience of its online shopping platform. The company employs Prometheus and Grafana for monitoring and alerting, which play a crucial role in SRE by providing real-time insights into the system's health and performance.

Continuous testing in this scenario is handled using Google's open source framework, Spinnaker, which supports advanced deployment strategies such as canary and blue/green deployments. This allows the SRE team to test new releases incrementally in live environments, thereby minimizing disruptions and reducing the risk of deploying faulty software to all users.

For maintaining quality and security, the company integrates SonarQube for static code analysis to maintain high code standards and detect security flaws early in the development process. Additionally, security is bolstered using tools such as HashiCorp Vault, which manages secrets and protects sensitive data across applications and infrastructure.

Feedback mechanisms are implemented using tools such as **Elasticsearch, Logstash, and Kibana (ELK)** stack to aggregate logs and facilitate detailed analysis of user interactions and system errors. This data is crucial for the SRE team to continuously improve system stability and performance based on actual user experiences and system behavior.

By integrating these tools into its SRE practices, the e-commerce company ensures that its platform remains reliable, secure, and performant, directly contributing to a better customer experience and increased business continuity.

Sustaining integrations

There are many use cases for continuous testing, continuous quality, continuous security, and continuous feedback for DevOps, DevSecOps, and SRE. Implementing and achieving working solutions for all of these use cases requires a considerable investment to transform practices for people (including culture), processes, and technologies, as illustrated in *Figure 9.5*.

Figure 9.5 – Sustaining integrations

However, after a successful transformation, the practices may erode and the gains may be impacted if practices are not put in place to sustain the people, processes, and tools for each of the use cases. This section describes recommended practices that will sustain the gains while allowing for ongoing improvements to the practices.

Sustaining the gains achieved through the transformation of DevOps, DevSecOps, and SRE practices into continuous methodologies for testing, quality, security, and feedback requires an ongoing commitment to nurturing people, refining processes, and leveraging technology effectively. To ensure that these gains are not only maintained but also enhanced over time, organizations must adopt practices that promote a culture of continuous improvement, adaptability, and resilience. The following recommendations are designed to sustain and further develop the successes achieved through the implementation of these practices:

- **Cultivating people and culture**:

  - **Foster a culture of continuous learning and improvement**: Encourage an environment where learning from failures is seen as an opportunity for growth. Provide regular training, workshops, and resources to keep teams up to date with the latest practices, tools, and technologies.

- **Promote ownership and accountability**: Empower teams by promoting a sense of ownership over their work. This includes responsibility for the quality, security, and reliability of the systems they develop and maintain.

- **Encourage collaboration and cross-functional teams**: Break down silos between development, operations, security, and other departments to encourage a holistic approach to system development and maintenance. Use cross-functional teams to foster a culture where different perspectives are valued and leveraged.

- **Recognize and reward contributions**: Implement recognition programs that acknowledge individual and team contributions to continuous improvement efforts. This motivates ongoing participation and engagement with the continuous practices.

- **Refining processes**:

  - **Implement continuous feedback loops**: Establish mechanisms for collecting, analyzing, and acting on feedback from all stakeholders, including end users, development teams, operations, and security. This should be an integral part of all stages of the life cycle, from requirements gathering to operations.

  - **Regularly review and update practices**: Conduct periodic reviews of the practices and processes to identify areas for improvement. Use metrics and KPIs to measure the effectiveness of practices and make data-driven decisions to refine them.

  - **Standardize and document best practices**: Develop and maintain a repository of best practices, guidelines, and lessons learned. Ensure that these resources are easily accessible to all team members to promote consistency and efficiency across projects.

  - **Plan for adaptability**: Design processes that are flexible and can easily adapt to changes in technology, market demands, and organizational goals. This includes embracing agile methodologies and being open to adjusting practices as needed.

- **Leveraging technology**:

  - **Invest in the right tools**: Choose tools that align with the organization's goals and practices. Ensure that they integrate well with existing systems and can scale to meet future needs.

  - **Automate wherever possible**: Automate repetitive and manual tasks to increase efficiency and reduce the likelihood of human error. This includes aspects of testing, deployment, monitoring, and feedback collection. Automation enables organizations to focus resources on other high-value tasks.

  - **Ensure tool compatibility and integration**: Tools should not only be selected based on individual capabilities but also on how well they integrate with other tools and systems within the organization. This facilitates smoother workflows and enhances collaboration across teams.

- **Regularly evaluate and update tools**: As part of the continuous improvement process, regularly assess the effectiveness of tools and technologies. Be prepared to replace or upgrade tools as better options become available or as the needs of the organization evolve.

Sustaining the gains from implementing continuous testing, quality, security, and feedback practices in DevOps, DevSecOps, and SRE requires a proactive approach to managing people, processes, and technology. By fostering a culture of continuous learning, collaboration, and improvement, refining processes to be adaptable and efficient, and thoughtfully selecting and managing technology tools, organizations can not only sustain their initial successes but also continuously enhance their practices over time. This ongoing commitment ensures that the organization remains resilient, competitive, and aligned with its strategic goals, even as it evolves and adapts to new challenges and opportunities.

Summary

This chapter described a comprehensive array of use cases for continuous testing, continuous quality, continuous security, and continuous feedback for DevOps, DevSecOps, and SRE practices. It explained how integrating these continuous practices enhances the efficiency, security, and reliability of continuous development, delivery, and operations. By delving into specific use cases at each stage of DevOps, DevSecOps, and SRE processes, the chapter provided insights into the practical application of these methodologies, their significance in achieving operational excellence, and the challenges that organizations may face during implementation.

The chapter explained the importance of a strategic approach to overcome implementation challenges. Strategies such as fostering a culture of continuous learning, encouraging collaboration across functions, and leveraging automation and state-of-the-art tools were discussed as key enablers for successfully embedding these practices into organizational workflows.

By illustrating the tangible benefits and potential pitfalls within each use case, this chapter provided practitioners with the knowledge and tools necessary to navigate the complexities of integrating continuous practices into their operations.

As organizations look to transition from understanding the use cases of continuous testing, quality, security, and feedback to practical implementation, the next chapter offers a strategic roadmap tailored for this purpose.

10

Building Roadmaps for Implementation

In this chapter, we explain how to create effective roadmaps for implementing continuous testing, quality, security, and feedback within your organization. The chapter guides you through the foundational steps necessary to define a tailored roadmap, ensuring that your digital transformation journey is both strategic and aligned with organizational goals.

By the end of this chapter, you will have gained valuable skills, including how to meticulously define a roadmap for an implementation journey, tailored to the organization's unique challenges and objectives; strategies for creating dedicated roadmaps focused on continuous testing, quality, security, and feedback—each essential for the development of robust, reliable, and user-centric products—and techniques for achieving alignment on the roadmap, ensuring that every team member is committed to the vision and works collaboratively toward the shared goals.

This chapter is not just about learning how to create a document; it's about envisioning a future where continuous improvement is ingrained in the fabric of your organization. Through practical insights and actionable advice, it prepares you to lead your team toward excellence with confidence and clarity.

In this chapter, we'll cover the following main topics:

- Introduction to strategic roadmaps
- Creating a roadmap
- Creating a future state value stream map
- Roadmap for continuous testing
- Roadmap for continuous quality
- Roadmap for continuous security
- Roadmap for continuous feedback
- Getting alignment on the roadmap

Let's get started!

Introduction to strategic roadmaps

Mastering continuous testing, quality, security, and feedback within organizations requires strategic planning. This journey, marked by a shift toward more agile, responsive, and quality-focused development practices, requires not just vision but strategic roadmaps and plans, each serving unique but complementary purposes in the organizational transformation journey.

The difference between a roadmap and a plan

At its core, a roadmap is a high-level, visual representation that outlines the major steps an organization intends to take to achieve its strategic goals, such as those related to continuous testing, quality, security, and feedback.

Unlike a detailed plan, a roadmap provides a bird's-eye view of activities for people, processes, and technologies and the key milestones toward transformation goals. A roadmap offers flexibility and allows for adjustments as the landscape evolves. Strategic roadmaps provide the big picture from which detailed plans can be formulated.

In contrast, a plan is a detailed document that outlines the specific actions, resource assignments, timelines, and responsibilities, for people, processes, and technologies, required to achieve the objectives identified in the roadmap. It delves into the operational aspects, providing a step-by-step guide to execution. Plans are tactical and detailed, offering a clear blueprint for day-to-day activities and short-term objectives.

The benefits of roadmaps

Implementing a strategic roadmap brings several benefits to organizations undergoing transformation:

- **Alignment**: Firstly, it ensures alignment across all levels of the organization by clearly communicating activities to drive the vision, objectives, and strategic priorities. This alignment is crucial for securing buy-in and fostering a shared sense of purpose.

- **Flexibility**: Secondly, roadmaps provide flexibility, allowing organizations to navigate the uncertainties of technological advancements and market changes with agility. They serve as a guiding light, keeping the organization focused on the long-term goals while accommodating shifts in strategy.

- **Resource allocation**: Thirdly, roadmaps facilitate better resource allocation by highlighting priority areas, ensuring that investments in time, personnel, and technology are made wisely and in service of the overarching goals.

The importance of a roadmap

A roadmap is crucial for several reasons:

- **Clarity and direction**: It provides a clear vision and direction, aligning stakeholders and ensuring everyone understands the objectives and the path forward.

- **Resource allocation**: It helps in the efficient allocation of resources, ensuring that investments in time, money, and effort are directed toward strategic priorities.

- **Risk management**: It allows organizations to anticipate risks and develop mitigation strategies, reducing the likelihood of project failures.

- **Change management**: It supports effective change management by outlining the impact on processes, people, and technology, and preparing the organization for transformation.

- **Performance tracking**: It enables the tracking of progress against objectives and **Key Performance Indicators (KPIs)**, allowing for course corrections as needed.

The perils of proceeding without a roadmap

Without a strategic roadmap, the transformation toward continuous testing and associated practices poses significant risks. Without a clear roadmap, organizations may find themselves without a coherent strategic direction, leading to misaligned efforts and wasted resources. The lack of a guiding strategic vision can result in initiatives that may not effectively contribute to the overall objectives, leading to inefficiencies and potential setbacks. Additionally, the absence of a roadmap can hinder the organization's ability to adapt to changes and challenges, as there is no strategic framework in place to guide decision-making and adjustments.

Without a roadmap, an organization may face several negative consequences:

- **Lack of alignment**: Without a clear plan, different parts of the organization may pursue conflicting priorities, leading to inefficiencies and wasted resources.

- **Resource misallocation**: Without prioritization and planning, resources may be spread too thinly or invested in low-impact initiatives.

- **Increased risks**: The absence of a risk management plan can leave the organization vulnerable to unforeseen challenges, potentially derailing the transformation.

- **Resistance to change**: Without a change management strategy, the organization may encounter significant resistance from employees and other stakeholders, hindering the adoption of new technologies and processes.

- **Inability to measure success**: Without defined KPIs and metrics, it becomes difficult to measure progress and demonstrate the value of digital transformation efforts.

In summary, while both roadmaps and plans play vital roles in the transformation process, they serve distinct purposes. The roadmap offers a strategic overview and flexibility, guiding the organization toward its long-term vision for continuous testing, quality, security, and feedback. In contrast, the plan provides the tactical details necessary for implementation. Together, they form a comprehensive approach to organizational transformation, with a roadmap ensuring strategic alignment and adaptability, and a plan detailing the path to execution. The benefits of adopting a strategic roadmap cannot be overstated, nor can the risks of proceeding without one. As organizations navigate the complexities of transformation, the strategic roadmap stands as an indispensable tool for success.

Best formats to represent the roadmap

The best format to represent a digital transformation roadmap can vary depending on the organization's needs and preferences. However, visual representation, such as a Gantt chart, dashboard, or infographic, is often effective. These formats can help stakeholders quickly understand timelines, dependencies, and progress. Tools such as Microsoft Excel, Project, Trello, or specialized roadmap software can facilitate creating, updating, and sharing the roadmap.

Figure 10.1 is an example roadmap represented as a Gantt chart. The format has the advantage that any level of detail can be added by simply adding horizontal lines for *epics* and vertical lines to represent *time frames* and *themes*.

TRANSFORMATION	xQyy			xQyy			xQyy			xQyy			xQyy			xQyy			xQyy			xQyy		
	M1	M2	M3	M1	M2	M3	M1	M2	M3	M1	M2	M3	M1	M2	M3	M1	M2	M3	M1	M2	M3	M1	M2	M3
Project Theme(s)			THEME 1								THEME 2							THEME 3						
People																								
Kickoff																								
Training																								
Work																								
Process																								
Design																								
Implement																								
Operation																								
Technologies																								
Selection																								
Commission																								
Application																								
Progress Tracking																								
Metrics																								
Milestone																								
Review																								

Figure 10.1 – Transformation roadmap template

In summary, a well-structured roadmap is essential for guiding a large organization through the complex process of digital transformation. It ensures alignment, efficient use of resources, risk mitigation, and the ability to measure success, thereby increasing the likelihood of achieving the desired outcomes.

Creating a roadmap

Creating a roadmap for digital transformation involves a series of structured steps, as indicated in *Figure 10.2*. It requires the involvement of various stakeholders from across the organization. This process is essential to ensure the transformation successfully addresses three key areas: people/culture, processes, and technology (which includes products, tools, and infrastructure elements).

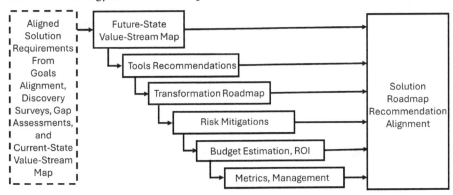

Figure 10.2 – Defining a roadmap for implementation

Let's break down the steps, identify the necessary participants, discuss how to evaluate alternatives, and determine when the roadmap is acceptable.

Steps to creating a roadmap

Following are the list of steps:

1. **Define vision and objectives**: As explained in *Chapter 5*, establish a clear vision for digital transformation and specific objectives. This should align with the organization's strategic goals. For example, establishing a capability maturity model, as explained in *Chapter 4*, can help determine capability levels that can identify the current level and goal level of capabilities.

2. **Conduct a current-state assessment and gap analysis**: As explained in *Chapter 6*, evaluate the current state of people (culture and capabilities), processes, and technology. Understand the maturity level, capabilities, and gaps in the current-state value stream. This is done using current-state surveys, gap assessments, and the **Current State Value Stream Map** (CSVSM). Identify the gaps between the current state and the desired future state. This analysis should cover skills, technology, processes, and cultural aspects.

3. **Identify priority areas and tools**: Determine key areas for digital transformation focus. Prioritization is crucial to tackle high-impact areas first. For this step, it is recommended to conduct a **Future State Value Stream Map** (FSVSM) as described later in this chapter. For each priority area, outline specific initiatives or projects, including objectives, required resources, and expected outcomes. It is recommended to use the methodology for selecting tool platforms and tools described in *Chapter 7*.

4. **Plan timeline and milestones**: Create a phased timeline with clear milestones. This should include short-term wins, long-term transformation goals, and the capability maturity model, as explained in *Chapters 4* and *5*.

5. **Identify risks and mitigation strategies**: Anticipate potential risks and develop strategies to address them.

6. **Allocate budget and resources**: Estimate the investment needed, including budgeting for technology acquisition, training, and hiring. Estimate the **Return on Investment** (**ROI**) as needed to justify the investment.

7. **Define success metrics and a change management plan**: Develop strategies for managing organizational change, focusing on communication, training, and support systems. Establish how success will be measured, using specific, measurable milestones and KPIs. Refer to *Chapter 12* for methodologies for establishing metrics.

Who should be involved

The following stakeholders should be involved and contribute to the roadmap:

- **Executive leadership**: Provides vision and support and ensures alignment with the organizational strategy.

- **Product development and QA leaders**: Provide insights from a product development and quality assurance perspective.

- **IT department**: Offers insights on current technology infrastructure and capabilities.

- **Human resources**: Facilitate the transformation of people and culture, including training and development.

- **Operational leaders**: Provide insights into current processes and help design future-state processes.

- **Digital transformation office or team (if available)**: Leads and coordinates the transformation efforts.

- **Finance department**: Assists with budgeting and financial planning.

- **External consultants (optional)**: Offer expertise and an outside perspective on digital transformation.

Evaluating roadmap alternatives

When creating a roadmap, it is normal to consider alternative solutions before deciding on one. The following are considerations for choosing a roadmap from a set of alternatives:

- **Alignment with strategic goals**: Ensure the roadmap aligns with the broader organizational objectives.

- **Feasibility and risk**: Assess the feasibility of each alternative and its associated risks.

- **Cost-benefit analysis**: Evaluate the expected benefits against the costs and resources required.

- **Stakeholder impact**: Consider the impact on employees, customers, and other key stakeholders.

- **Flexibility**: Ensure the roadmap is adaptable to changing circumstances and technologies.

Determining an acceptable roadmap

An acceptable roadmap is one that is as follows:

- **Aligns with strategic objectives**: Clearly supports the organization's strategic goals and objectives.

- **Is realistic and feasible**: Can realistically be implemented within the given constraints of time, budget, and resources.

- **Has stakeholder buy-in**: Is supported by key stakeholders across the organization.

- **Addresses people, processes, and technology**: Provides a comprehensive transformation across all three domains.

- **Includes clear metrics for success**: Features well-defined KPIs that allow for tracking progress and measuring success.

Once you have a roadmap that meets these criteria and has been reviewed and approved by key stakeholders, you can consider it acceptable and begin implementation. Regular reviews and adjustments will likely be necessary as the transformation progresses, ensuring the roadmap remains aligned with organizational goals and adapts to any changes in the external environment or organizational priorities.

Creating a future state value stream map (FSVSM)

An FSVSM is a visual tool used in lean management that outlines the desired future state of processes within an organization, focusing on value delivery from start to finish. Unlike traditional value-stream mapping, which documents current processes, inefficiencies, and waste, the FSVSM depicts how processes should ideally operate in the future after improvements are implemented. It aims to streamline operations, enhance productivity, and eliminate non-value-added activities, thereby ensuring that every step in the process directly contributes to the final output's value.

The importance of FSVSMs in establishing transformation roadmaps

The following are reasons explaining why an FSVSM is important to establish a transformation roadmap:

- **Vision clarification**: FSVSMs help articulate a clear and shared vision of the desired future state of operations. By visualizing the end goal, organizations can ensure that all stakeholders are aligned and motivated toward the same objectives.

- **Gap identification**: Mapping the future state allows organizations to identify the gaps between current processes and the optimal future operations. Understanding these gaps is crucial for developing targeted strategies to bridge them, thereby making the transformation roadmap more effective.

- **Prioritization of initiatives**: FSVSMs facilitate the identification of high-impact areas for improvement. Organizations can use this information to prioritize initiatives, allocate resources efficiently, and focus on changes that offer the most significant benefits in terms of value delivery and waste reduction.

- **Enhanced collaboration and communication**: The visual and collaborative nature of FSVSMs fosters better communication among team members, departments, and stakeholders. This improved communication is essential for coordinating efforts, sharing best practices, and ensuring that the transformation is comprehensive and cohesive.

- **Continuous improvement culture**: By regularly updating the FSVSM to reflect achieved improvements and setting new targets, organizations can foster a culture of continuous improvement. This iterative process encourages constant evaluation and adaptation, which is vital in today's fast-paced and ever-changing business environment.

- **Strategic planning support**: FSVSMs provide a structured approach to strategic planning by outlining the steps necessary to achieve the desired future state. They serve as a foundation for developing detailed action plans, setting realistic timelines, and measuring progress toward the transformation goals.

In conclusion, FSVSMs are indispensable tools in the planning and execution of organizational transformations. They offer a strategic vision that guides the development of effective transformation roadmaps, ensuring that efforts are focused, aligned, and driven by value. By leveraging FSVSMs, organizations can navigate the complexities of change with clarity and precision, ultimately achieving their goals of improved efficiency, quality, and customer satisfaction.

FSVSM workshop

Creating an FSVSM for a digital transformation involves several key steps and preparatory actions to ensure the mapping process is both efficient and effective. Here's a brief outline of the process and the importance of various aspects:

1. **Set a specific goal for the value stream**: Setting a specific goal is crucial because it aligns with the focus of the digital transformation efforts. It helps in identifying which processes need improvement and ensures that the future-state map is designed with clear objectives in mind, such as reducing waste, improving process flow, or enhancing customer satisfaction.

2. **Prepare participants in advance**: Before conducting the FSVSM workshop, it's essential to prepare the participants. This involves selecting a cross-functional team and providing them with background information on value-stream mapping, the objectives of the workshop, and any pre-reading materials or data they might need. Preparation ensures that participants come informed and ready to contribute meaningfully to the mapping process.

3. **Review the current-state mapping**: Begin by reviewing the current state of the value stream to understand the existing processes, information flows, and pain points. This step involves documenting steps, delays, and inefficiencies to create a baseline for improvement.

4. **Identify bottlenecks across the value stream**: Through the current state mapping, identify bottlenecks and areas of waste. Bottlenecks are critical because they slow down the process flow and can significantly impact overall performance. Identifying these allows for targeted interventions in the future-state design.

5. **Design the future state map**: Leveraging insights from the current-state analysis and keeping the specific goal in mind, design the future state of the value stream. This involves reimagining processes with digital technologies, eliminating non-value-added steps, and resolving identified bottlenecks to create a more efficient and effective value stream.

6. **Develop an implementation roadmap**: Develop a roadmap to transition from the current state to the future state. The roadmap should include specific projects or initiatives, resources required, timelines, and metrics for measuring success. The roadmap should include strategic changes for people, processes, and technologies needed to realize the transformation goals.

7. **Summarize the outputs of the workshop**: The primary outputs include a detailed FSVSM, a comprehensive implementation roadmap, identified areas for immediate improvement, and a list of potential digital technologies or solutions to be adopted. Additionally, there should be a clear set of performance metrics to track progress toward the future state.

The process of creating an FSVSM for digital transformation is fundamental in ensuring that efforts are directed toward the most impactful areas. By setting a specific goal, preparing participants, identifying bottlenecks, and thoroughly planning for implementation, organizations can significantly improve their operations, enhance customer value, and achieve a competitive advantage in the digital age.

Next, let's shift our focus to the specifics of integrating continuous testing into the software development life cycle.

Roadmap for continuous testing

This section outlines strategies for ensuring that testing is an ongoing process, allowing for early detection of issues and ensuring that quality is built into the product from the start.

Determining an effective roadmap for continuous testing requires careful consideration of the results of the discovery survey, capability maturity assessment, gap assessment, tools assessments, and CSVSM, as explained in previous chapters.

There is no standard roadmap because every organization and application has unique current state and future state priorities for continuous testing.

Figure 10.3 is an example of a roadmap for continuous testing. In this example, the organization's current state of continuous testing was assessed to be capability level 2 (i.e., continuous integration) and the desired goal for the organization is to transform to level 4 (i.e., continuous deployment).

Continuous Testing TRANSFORMATION	xQyy			xQyy			xQyy			xQyy			xQyy			xQyy
	M1	M2	M3	M1	M2	M3	M1	M2	M3	M1	M2	M3	M1	M2	M3	M1
Project Theme(s)				P1: L2 to L3:Continuous Delivery						P2: L3 to L4: Continuous Deployment						
People																
Kickoff	K1									K2						
Training		T1						T2			T3					T4
Work				Roles and RACIs for CICD test auto							Deployment Roles and RACIs					
Process																
Design		D1								D2						
Implement				Automate Test Environments							BG Env,					
Implement				Auto-Testing in CD							Blue-Green Tests					
Operation							CD test coverage							Trial		
Technologies																
Selection	S1									S2						
Commission		C1									C2					
Application				CD Test Automation using selected tools							CD2 tools in use					
Progress Tracking																
Dashboard			Dashboard CT Metrics P1							Dashboard CT Metrics P2						
Milestone		M1			M2			M3				M4				M5
Review				R1						R2				R3		R4

Figure 10.3 – Roadmap for continuous testing (example)

The roadmap has two project themes:

- **Theme P1**, to transform continuous testing practices from level 2, continuous integration, to level 3, continuous delivery.

- **Theme P2**, to transform from level 3 to level 4, continuous deployment.

The roadmap for theme P1 includes the following Epics to transform people, processes, technologies, and progress tracking:

- **People**: A kick-off meeting to get team alignment on implementation, training, and the roles and responsibility matrix (i.e., RACI) for CI/CD test automation.

- **Processes**: Design activities, automation of test environments, automation of testing for continuous delivery, and validation of CD test coverage.

- **Technologies**: Tool selection, commissioning new tools, and the use of the selected tools for CD automation.

- **Progress tracking**: Defining dashboard metrics and milestones and reviewing checkpoints for continuous testing.

The roadmap for theme P2 includes the following Epics to transform people, processes, technologies, and progress tracking:

- **People**: A kickoff meeting to get team alignment on implementation, training, and updated roles and RACIs for CI/CD test automation.

- **Processes**: Process design activities, blue-green environment process automation, and verification and trial.

- **Technologies**: Tool selection for deployment, commission of new tools, and use of the selected tools for deployment automation.

- **Progress tracking**: Update dashboard metrics and milestones and review continuous testing checkpoints for deployment automation.

This roadmap is a high-level approach upon which detailed implementation plans for continuous testing can be based, including stories and tasks.

Roadmap for continuous quality

Continuous quality goes beyond testing, encompassing all aspects of the product life cycle to ensure excellence. This section discusses how to create a culture of quality where every team member is an advocate for excellence, leveraging tools, processes, and metrics to maintain high standards.

There is no standard roadmap because every organization and application has unique current-state and future-state priorities for continuous quality.

Figure 10.4 is an example of a roadmap for continuous quality. In this example, the organization's current state of continuous quality was assessed to be a level 2 capability, continuous integration, and the desired goal for the organization is to transform to level 4, continuous deployment.

Continuous Quality TRANSFORMATION	xQyy			xQyy			xQyy			xQyy			xQyy			
	M1	M2	M3	M1	M2	M3	M1	M2	M3	M1	M2	M3	M1	M2	M3	M1
Project Theme(s)				P1: Continuous Delivery CD1 Quality L3									P2: Cont. Deploy CD2 Quality L4			
People																
Kickoff	K1									K2						
Training		T1							T2		T3					T4
Work				Roles and RACIs for CICD quality							CD2 Quality Roles and RACIs					
Process																
Design		D1								D2						
Implement		Automate CD Quality Checks									CD2 Quality checks					
Operation							CD Quality Checks Trial							Trial		
Technologies																
Selection		S1								S2						
Commission			C1								C2					
Application			CD1 Qualiy Automation using selected tools										CD2 Quality tools in use			
Progress Tracking																
Dashboard				Dashboard CQ Metrics P1							Dashboard CQ Metrics P2					
Milestone		M1			M2				M3				M4			M5
Review					R1					R2				R3		R4

Figure 10.4 – Roadmap for continuous quality (example)

The roadmap has two project themes:

- **Theme P1**, to transform continuous quality practices from level 2, continuous integration, to level 3, continuous delivery.

- **Theme P2**, to transform continuous quality practices from level 3 to level 4, continuous deployment.

The roadmap for theme P1 includes the following Epics to transform people, processes, technologies, and progress tracking:

- **People**: A kickoff meeting to get team alignment on implementation, training, and roles and RACIs for the CI/CD quality plan.

- **Processes**: Design activities, automation of continuous quality checks for continuous delivery, and validation of continuous delivery quality checks.

- **Technologies**: Selection of quality-checking tools for continuous delivery, commission of new tools, and use of the selected tools for continuous delivery quality automation.

- **Progress tracking**: Define dashboard metrics and milestones and review checkpoints for continuous quality.

The roadmap for theme P2 includes the following Epics to transform people, processes, technologies, and progress tracking:

- **People**: A kickoff meeting to get team alignment on implementation, training, and updated roles and RACIs for CI/CD quality automation.

- **Processes**: Process design activities, P2 process automation, verification, and trial.

- **Technologies**: Selection of quality checking tools for deployment, commission of new tools, and use of the selected quality checking tools for deployment automation.

- **Progress tracking**: Update dashboard metrics and milestones and review checkpoints for continuous quality for deployment.

This roadmap is a high-level approach for continuous quality, upon which detailed implementation plans can be built, including stories and tasks.

Roadmap for continuous security

With cybersecurity threats on the rise, this section highlights the necessity of embedding security practices into every stage of development. It provides a blueprint for a proactive security posture, ensuring that products are not only functional but also secure against potential threats.

There is no standard roadmap because every organization and application has unique current state and future state priorities for continuous security.

Figure 10.5 is an example of a roadmap for continuous security. In this example, the organization's current state of continuous security was assessed to be a level 2 capability, continuous integration, and the desired goal for the organization is to transform to level 4, continuous deployment.

Continuous Security TRANSFORMATION	xQyy			xQyy			xQyy			xQyy			xQyy			
	M1	M2	M3	M1	M2	M3	M1	M2	M3	M1	M2	M3	M1	M2	M3	M1
Project Theme(s)	P1: Continuous Delivery CD1 Security L3									P2: Cont. Deploy CD2 Security L4						
People																
Kickoff	K1									K2						
Training		T1							T2		T3					T4
Work		Roles and RACIs for CICD security									CD2 Security Roles and RACIs					
Process																
Design			D1								D2					
Implement		Automate CD Security Checks									CD2 Security checks					
Operation						CD Security Checks Trial								Trial		
Technologies																
Selection		S1									S2					
Commission			C1								C2					
Application			CD1 Security Automation using selected tools									CD2 Security tools used				
Progress Tracking																
Dashboard			Dashboard CS Metrics P1								Dashboard CS Metrics P2					
Milestone		M1			M2				M3				M4			M5
Review				R1						R2				R3		R4

Figure 10.5 – Roadmap for continuous security (example)

The roadmap has two project themes:

- **Theme P1**, to transform continuous security practices from level 2, continuous integration, to level 3, continuous delivery.

- **Theme P2**, to transform continuous security practices from level 3 to level 4, continuous deployment.

The roadmap for theme P1 includes the following Epics to transform people, processes, technologies, and progress tracking:

- **People**: A kickoff meeting to get team alignment on implementation, training, and roles and RACIs for the CI/CD security strategy.

- **Processes**: Design activities, automation of continuous quality checks for continuous delivery, and validation of continuous delivery security checks.

- **Technologies**: Selection of quality checking tools for continuous delivery, commission of new tools, and the use of the selected tools for security automation for continuous delivery.

- **Progress tracking**: Define dashboard metrics and milestones and review checkpoints for continuous security.

The roadmap for theme P2 includes the following Epics to transform people, processes, technologies, and progress tracking:

- **People**: A kickoff meeting to get team alignment on implementation, training, and updated roles and RACIs for security automation.

- **Processes**: Process design activities, P2 process automation, verification, and trial.

- **Technologies**: Selection of security checking tools for deployment, commission of new tools, and use of the selected quality checking tools for deployment automation.

- **Progress tracking**: Update dashboard metrics and milestones and review checkpoints for continuous security for deployment.

This roadmap is a high-level approach for continuous security, upon which detailed implementation plans can be built, including stories and tasks.

Roadmap for continuous feedback

This part underscores the value of feedback in the iterative development process. It shows how to establish channels for continuous feedback from users and stakeholders, facilitating rapid adjustments and improvements to meet their needs effectively.

There is no standard roadmap because every organization and application has unique current-state and future-state priorities for continuous feedback.

Figure 10.6 is an example of a roadmap for continuous feedback. In this example, the organization's current state of continuous feedback was assessed to be a level 2 capability, continuous integration, and the desired goal for the organization is to transform to level 4, continuous deployment.

Continuous Feedback TRANSFORMATION		xQyy			xQyy			xQyy			xQyy			xQyy			
		M1	M2	M3	M1	M2	M3	M1	M2	M3	M1	M2	M3	M1	M2	M3	M1
Project Theme(s)		P1: Continuous Delivery CD1 Feedback L3									P2: Cont. Deploy CD2 Feedback L4						
People																	
	Kickoff	K1									K2						
	Training		T1							T2		T3					T4
	Work			Roles and RACIs for CICD Feedback							CD2 Feedback Roles and RACIs						
Process																	
	Design		D1								D2						
	Implement			Automate CD Feedback							CD2 Feedback						
	Operation							CD Feedback Trial							Trial		
Technologies																	
	Selection	S1									S2						
	Commission			C1								C2					
	Application				CD1 Feedback Automation							CD2 Feedback tools use					
Progress Tracking																	
	Dashboard			Dashboard CF Metrics P1							Dashboard CF Metrics P2						
	Milestone		M1			M2			M3				M4				M5
	Review				R1					R2					R3		R4

Figure 10.6 – Roadmap for continuous feedback (example)

The roadmap has two project themes:

- **Theme P1**, to transform continuous feedback practices from level 2, continuous integration, to level 3, continuous delivery.
- **Theme P2**, to transform continuous feedback practices from level 3 to level 4, continuous deployment.

The roadmap for theme P1 includes the following Epics to transform people, processes, technologies, and progress tracking:

- **People**: A kickoff meeting to get team alignment on implementation, training, and roles and RACIs for feedback strategy.
- **Processes**: Design activities, automation of continuous feedback for continuous delivery, and validation of feedback for continuous delivery.
- **Technologies**: Selection of feedback tools for continuous delivery, commission of new tools, and use of the selected tools for feedback automation for continuous delivery.
- **Progress tracking**: Define dashboard metrics and milestones and review checkpoints for continuous feedback.

The roadmap for theme P2 includes the following Epics to transform people, processes, technologies, and progress tracking:

- **People**: A kickoff meeting to get team alignment on implementation, training, and updated roles and RACIs for CI/CD feedback.
- **Processes**: Process design activities, P2 process automation, verification, and trial.
- **Technologies**: Selection of feedback tools for deployment, commission of new tools, and use of the selected feedback tools for deployment automation.
- **Progress tracking**: Update dashboard metrics and milestones and review checkpoints for continuous feedback for deployment.

This roadmap is a high-level approach for continuous feedback, upon which detailed implementation plans can be built, including stories and tasks.

Alignment on the roadmap

This section addresses one of the most crucial aspects of roadmap implementation—securing alignment across all levels of the organization. It offers practical advice on engaging stakeholders, fostering collaboration, and ensuring that the roadmap has the support it needs to succeed.

It is recommended to develop distinct roadmaps for continuous testing, quality, security, and feedback, to ensure the distinct goals for each are considered. It is important to obtain consensus from the transformation team for all themes and epics. Once there is agreement, then it is advisable to develop a combined roadmap to simplify program tracking for the whole set of themes and Epics.

Identifying risks and mitigation strategies

Transforming an organization toward a more mature capability in continuous testing, quality, security, and feedback involves significant changes in processes, technology, and culture. Recognizing and mitigating the risks associated with these changes is crucial for a successful transition, as illustrated in *Figure 10.7*:

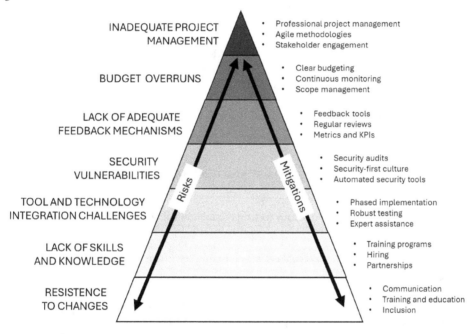

Figure 10.7 – Risks and mitigations

Here, the risk and mitigation strategies are explained:

- **Resistance to change**: Employees may resist new methodologies, tools, and changes in their routines, especially if they perceive them as threats to their current jobs or skills.

  - **Mitigation strategies**:

    - **Communication**: Regularly communicate the benefits and reasons behind the changes to all stakeholders.

- **Training and education**: Provide comprehensive training to help employees adapt to new tools and practices.

- **Inclusion**: Involve employees in the transition process through feedback sessions and pilot programs.

- **Inadequate skills and knowledge**: The current workforce may lack the necessary skills to implement and maintain new technologies and practices effectively:

  - **Mitigation strategies**:

    - **Training programs**: Invest in training programs and workshops to upskill employees.

    - **Hiring**: Bring in new talent with the necessary skills or experience in areas such as DevOps, cybersecurity, and automated testing.

    - **Partnerships**: Collaborate with external consultants or firms that specialize in digital transformations.

- **Tool and technology integration challenges**: Integrating new tools with existing systems can be complex and disruptive, potentially leading to downtime or data integrity issues:

  - **Mitigation strategies**:

    - **Phased implementation**: Roll out new tools and processes gradually to manage integration complexity.

    - **Robust testing**: Employ rigorous testing of new tools in a controlled environment before full-scale implementation.

    - **Expert assistance**: Use external experts for complex integrations to ensure best practices are followed.

- **Security vulnerabilities**: New tools and more frequent deployments can introduce security vulnerabilities, especially if not properly managed:

  - **Mitigation strategies**:

    - **Security audits**: Conduct regular security audits and vulnerability assessments.

    - **Security-first culture**: Foster a culture of security awareness and ensure security practices are integrated into the development process from the start.

    - **Automated security tools**: Implement automated security testing tools to continuously scan for vulnerabilities.

- **Lack of adequate feedback mechanisms**: Without effective feedback mechanisms, continuous improvements in quality and performance can stall:

  - **Mitigation strategies**:

    - **Feedback tools**: Implement tools that facilitate real-time feedback from end users and stakeholders.

    - **Regular reviews**: Schedule regular review meetings to discuss feedback and incorporate it into development cycles.

    - **Metrics and KPIs**: Define and track KPIs to measure the impact of feedback on product quality and team performance.

- **Budget overruns**: Transformations can become significantly more expensive than anticipated due to unforeseen challenges or scope creep:

  - **Mitigation strategies**:

    - **Clear budgeting**: Establish clear budget constraints and approval processes for expenditures related to the transformation.

    - **Continuous monitoring**: Regularly monitor and review expenses and progress against set goals to ensure alignment.

    - **Scope management**: Clearly define the scope and objectives of the transformation to avoid expansion beyond original intentions.

- **Inadequate project management**: Poor project management can lead to misaligned goals, missed deadlines, and project failure:

  - **Mitigation strategies**:

    - **Professional project management**: Employ experienced project managers who are familiar with IT transformations.

    - **Agile methodologies**: Use agile methodologies to manage projects, allowing for flexibility and regular reassessment of project direction and priorities.

    - **Stakeholder engagement**: Keep all stakeholders engaged and informed throughout the project life cycle to ensure alignment and address issues promptly.

By carefully planning and addressing these risks, an organization can successfully navigate the complexities of adopting continuous testing, quality, security, and feedback, leading to a more resilient and efficient operational model.

Allocating budget and resources

Determining the implementation budget and projecting the ROI for transformations in continuous testing, quality, security, and feedback requires a comprehensive approach. Here are recommended steps to establish a budget and estimate the ROI, including considerations for technology acquisition, training, hiring, and process automation:

1. **Clarify the scope and objectives**:

    *   **Identify specific goals**: Clearly define what the transformation intends to achieve, including specific improvements to quality, speed, security, and feedback mechanisms.

    *   **Requirement analysis**: Assess current capabilities and identify gaps that the transformation needs to address. This includes technology, skills, processes, and cultural aspects.

2. **Estimate the costs**:

    *   **Technology acquisition**: Estimate costs for purchasing new tools and integrating them with existing systems. This might include software licenses, hardware, and cloud services.

    *   **Training**: Allocate budget for training current employees on new tools, methodologies, and best practices.

    *   **Hiring**: Determine costs associated with recruiting new staff with the necessary expertise if current employees cannot fill all roles.

    *   **Process automation**: Assess costs for automating processes, which may involve software development, buying off-the-shelf solutions, or customizing existing tools.

    *   **Infrastructure upgrades**: Include potential upgrades to infrastructure that might be necessary to support new technologies and increased loads.

    *   **Consultancy and external expertise**: Budget for external consultants if specialized knowledge is required for implementation.

3. **Calculate the ROI**:

    *   **Quantify benefits**: Estimate tangible benefits such as reduced time to market, improved product quality, decreased downtime, and lower security incident costs.

    *   **Indirect benefits**: Consider indirect benefits such as improved customer satisfaction, enhanced brand reputation, and increased employee satisfaction.

    *   **Cost savings**: Calculate cost savings from reduced error rates, fewer security breaches, and decreased need for manual testing.

- **ROI formula**: Use the formula:

  - %ROI={(Net Benefits–Total Cost)/(Total Cost)}×100%, where net benefits include both revenue increases and cost savings.

4. **Allocate a budget**:

   - **Priority-based funding**: Allocate funds based on the prioritized needs identified in the scope analysis. Critical areas that drive the biggest ROI should be prioritized.

   - **Phased allocation**: Consider a phased approach to budget allocation, releasing funds as specific milestones are achieved to keep the project on track and within budget.

   - **Contingency fund**: Set aside a portion of the budget (typically 5-10%) for unforeseen costs or adjustments in the transformation process.

5. **Monitor and adjust**:

   - **Performance metrics**: Establish KPIs to monitor the success of the implementation across different stages.

   - **Feedback loops**: Integrate feedback mechanisms to continuously improve and adjust the transformation strategy as needed.

   - **Budget review**: Regularly review the budget against the actual spending and adjust forecasts based on project progress and external factors.

6. **Report and communicate**:

   - **Stakeholder reports**: Keep all stakeholders informed with regular updates on budget usage, ROI projections, and project status.

   - **Transparent communication**: Ensure that all financial decisions and adjustments are transparently communicated to involved parties to maintain trust and alignment.

This strategic approach will help in not only setting up a realistic budget for the transformation but also in calculating a reliable ROI that can justify the investments. Moreover, ongoing monitoring and adaptive management of the budget and resources will ensure that the transformation remains aligned with organizational goals and market demands.

Defining success metrics and a change management plan

For a successful transformation toward more mature capabilities in continuous testing, quality, security, and feedback, it's essential to define clear success metrics and establish a robust change management plan. This will help ensure that the transformation is effectively managed and aligned with the organizational goals. Here's a detailed plan.

Success metrics

The most popular metrics for measuring success are the de facto industry-standard DORA metrics. The following list provides more details on these metrics, as well as some information on additional metrics:

- **DORA metrics**:

  - **Deployment frequency**: Measure how often deployments occur. A higher frequency typically indicates more mature continuous testing and deployment practices.

  - **Lead time for changes**: Track the time it takes from code commit to code successfully running in production. Shorter lead times are a sign of efficient processes.

  - **Change failure rate**: Monitor the percentage of deployments causing a failure in production. Lower failure rates are indicative of higher quality and better testing.

  - **Time to restore service**: Measure the time it takes to recover from a failure in production. Faster recovery times demonstrate better incident management and reliability.

- **Additional metrics**:

  - **Automated test coverage**: Percentage of code covered by automated tests, aiming for high coverage to ensure quality.

  - **Bug detection rate**: The rate at which bugs are detected and fixed during development phases.

  - **Security incident rate**: Track the frequency and severity of security incidents to gauge improvements in security measures.

  - **User feedback response time**: Time taken to respond to and address user feedback, reflecting on the feedback loop effectiveness.

  - **Employee satisfaction**: Measure changes in employee satisfaction regarding new processes and tools, as it impacts adoption and productivity.

Change management plan

The following are good practices for managing the changes for the transformation:

- **Communication plan**:

  - **Kickoff meetings**: Launch the transformation with meetings that outline goals, expectations, and timelines to all stakeholders.

  - **Regular updates**: Provide ongoing updates through emails, newsletters, or dedicated intranet sections to keep all parties informed about progress and changes.

  - **Feedback channels**: Establish open channels (such as surveys, forums, and meetings) for employees to voice concerns and provide feedback on the transformation process.

- **Training plan**:

  - **Initial training sessions**: Conduct comprehensive training for all impacted employees on new tools, processes, and best practices at the start of the transformation.

  - **Ongoing education**: Offer continuous learning opportunities through workshops, webinars, and e-learning courses to adapt to evolving tools and methods.

  - **Cross-functional training**: Promote understanding across different roles and departments to foster collaboration and comprehension of the entire life cycle.

- **Support systems**:

  - **Service desks and resource centers**: Set up service desks to address technical questions and resource centers where employees can find documentation and toolkits.

  - **Change agents and champions**: Identify and empower change champions within teams who can provide peer support and promote the benefits of new practices.

- **Measuring success**:

  - **Milestone reviews**: Set specific milestones, such as "Deploy automation tools across all development teams within six months" or "Achieve 75% automated test coverage within one year." Review progress in milestone meetings.

  - **Performance dashboards**: Use dashboards to visually monitor KPIs and metrics in real time, providing both high-level and detailed views of ongoing performance.

  - **Continuous improvement sessions**: Regularly schedule sessions to review successes and failures and to plan adjustments in strategy or focus as needed.

Establishing KPIs

The following are recommended practices to establish KPIs:

- **Deployment frequency**: Aim for daily deployments.
- **Lead time for changes**: Target reduction from months to weeks, or weeks to days.
- **Change failure rate**: Reduce from baseline by 50% within the first year.
- **Time to restore service**: Aim to cut recovery time by 75% within the first year.
- **Automated test coverage**: Increase coverage by 25% in the first six months.
- **Security incident rate**: Reduce incidents by 50% within the first year.

The combination of clearly defined metrics, an extensive change management plan, and constant evaluation will guide the organization toward successful implementation and sustainable transformation.

Summary

This chapter served as an essential guide for organizations looking to implement continuous testing, quality, security, and feedback within their digital transformation initiatives. The chapter began by outlining the importance of a well-defined, tailored roadmap, emphasizing the alignment of the roadmap with the organization's strategic objectives. It covered the foundational steps necessary for crafting an actionable plan that not only guides the implementation but also ensures it integrates smoothly with existing organizational structures and goals.

The middle sections of the chapter were divided into specific areas of focus: continuous testing, continuous quality, continuous security, and continuous feedback. Each section provided detailed strategies and frameworks for integrating these elements into the software development life cycle. This included advice on fostering a culture of quality and security, leveraging tools and processes to maintain high standards, and establishing channels for ongoing feedback from users and stakeholders to continuously refine the development process.

Toward the end of the chapter, there was a significant focus on alignment within the organization, which is critical for the successful implementation of the roadmap. This included practical advice on securing stakeholder buy-in and fostering an environment of collaboration across all levels of the organization. The chapter stressed the importance of communication, training, and proper support systems to assist employees in adapting to new technologies and methodologies.

Finally, the chapter wrapped up with insights into measuring the success of the implementation using specific, measurable KPIs and DORA metrics, such as deployment frequency and change failure rate. These metrics help in tracking progress and ensuring the transformation aligns with the desired outcomes, providing a clear view of the roadmap's effectiveness in real time. This structured approach not only facilitates a smooth transition but also maximizes the chances of achieving the targeted benefits of digital transformation efforts.

The next chapter will explore organization implementation patterns for the transformation to more mature capabilities for continuous testing, quality, security, and feedback.

11

Understanding Transformation Implementation Patterns

As organizations seek to streamline their projects and optimize team structures, the selection of effective implementation patterns becomes critical. This chapter dives into the world of implementation patterns, which are structured approaches that are proven to enhance the deployment and success of strategic roadmaps for organizations that wish to improve their capabilities for continuous testing, quality, security, and feedback. Understanding these patterns equips project managers and team leads with the knowledge to choose the best strategies tailored to their specific goals and challenges. We will explore what an implementation pattern is and detail four distinct and successful patterns in this chapter. As we delve into these patterns, we will also discuss which strategies have historically not succeeded and why these patterns should be avoided. Lastly, guidance on selecting the most appropriate implementation pattern will be provided, helping you apply practical criteria and considerations to your decision-making process.

By the end of this chapter, you will not only be able to identify and understand various implementation patterns but also be equipped with the skills to select and apply them effectively in their organizational context, ensuring a more robust and successful roadmap implementation.

In this chapter, we'll cover the following main topics:

- What is a transformation implementation pattern?
- Understanding transformation implementation patterns
- Patterns to avoid during implementation
- Selecting an implementation pattern

Let's get started!

What is a transformation implementation pattern?

We will begin by defining what exactly an implementation pattern is and why it is crucial for successful project execution. An implementation pattern can be seen as a blueprint or a formula that, when followed, significantly increases the likelihood of achieving project objectives smoothly and efficiently. This foundational knowledge sets the stage for deeper insights into each pattern's specific characteristics and advantages.

A transformation implementation pattern is a structured approach that organizations use to evolve from their current operational methodologies to a more advanced state, particularly in the domains of continuous testing, quality, security, and feedback. These patterns serve as detailed guides or frameworks designed to help organizations systematically improve their processes and tools, thus enhancing overall efficiency and performance. The goal of these patterns is not just to implement new tools or practices, but to fundamentally alter the way teams collaborate and function, ensuring that continuous improvement becomes an integral part of the organizational culture.

The essence of a transformation implementation pattern lies in its structured approach to change. It breaks down the transformation process into manageable, actionable steps, providing a clear roadmap that organizations can follow. This is particularly crucial when dealing with complex areas such as continuous testing and security, where multiple variables, including technology integration, process updates, and personnel training, must be harmoniously aligned. By following a transformation implementation pattern, organizations can tackle these challenges in a systematic manner, reducing risks and avoiding common pitfalls that could derail their progress.

Key components of effective implementation patterns

An effective transformation implementation pattern typically includes several critical components, as shown in *Figure 11.1*.

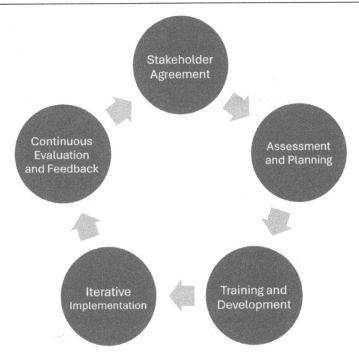

Figure 11.1 – The components of transformation implementation patterns

Let's have a look at the details:

1. **Stakeholder agreement**: Successful transformation requires buy-in from all levels of the organization. This component focuses on engaging with stakeholders to explain the benefits of the transformation, address concerns, and gather input. Effective stakeholder engagement ensures that the transformation is aligned with business objectives and enjoys broad support.

2. **Assessment and planning**: This stage involves a thorough assessment of the current state of the organization's testing, quality, security, and feedback mechanisms. The results of this assessment are then used to create a detailed plan that outlines the specific goals of the transformation and the steps needed to achieve them.

3. **Training and development**: As processes and tools are updated, personnel must be trained not only in the new systems but also in the principles of continuous improvement. This training ensures that team members can effectively utilize new tools and practices and are prepared to contribute to ongoing process enhancements.

4. **Iterative implementation**: Rather than attempting to overhaul systems all at once, the best implementation patterns advocate for an iterative approach. This allows organizations to test new methods on a small scale, refine them based on real-world feedback, and gradually expand successful practices across the entire organization.

5. **Continuous evaluation and feedback**: An integral part of any implementation pattern is the continuous evaluation of how the transformation is progressing. This involves regular check-ins, metrics analysis, and feedback sessions that help teams understand what is working and what needs adjustment.

While all transformation patterns have similar components, as described in this section, there are different patterns that suit different organizations and products. The next section describes an approach to choosing the best pattern for a particular organization.

Choosing the right pattern

Choosing the right transformation implementation pattern is critical because each organization has unique challenges, goals, and resources. Some patterns may focus more on technological integration, while others might prioritize cultural change or process reengineering. The choice of pattern should be based on a careful analysis of the organization's specific needs and the pattern's proven effectiveness in similar contexts.

For instance, an organization with a robust IT infrastructure but weak process discipline might benefit more from a pattern that emphasizes process standardization and training. Conversely, a company with strong operational processes but outdated technology might need a pattern that focuses on technological transformation.

The importance of selecting the right pattern

The end results of a transformation initiative can vary significantly based on the implementation pattern chosen. Selecting the right pattern can lead to a smoother transformation, higher team adoption rates, and more substantial improvements in testing, quality, security, and feedback mechanisms. On the other hand, choosing a mismatched pattern can result in wasted resources, low morale, and limited improvements, which might leave the organization no better off than before or possibly even in a worse state.

Therefore, it's crucial to consider the organization's current capabilities, the specific goals of the transformation, and the proven effectiveness of the available patterns. With careful selection and implementation, organizations can achieve a higher level of operational maturity, ensuring that they are better equipped to meet the demands of an increasingly complex and fast-paced business environment.

Understanding transformation implementation patterns

The subsequent sections of this chapter are dedicated to exploring four effective implementation patterns.

Dedicated platform team

The **dedicated platform team** structure emphasizes the benefits of having a centralized team focused solely on the platform's development and maintenance.

A dedicated platform team, as shown in *Figure 11.2*, is composed of experts in software development, testing, security, and operations who are tasked with the continuous improvement of the platform's infrastructure.

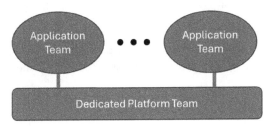

Advantages
- Specialized Focus and Expertise
- Enhanced Quality and Performance
- Faster Response Times
- Better Alignment with Business Goals
- Increased Scalability

Disadvantages
- Resource Intensity
- Potential Isolation
- Dependency and Bottlenecks
- Adaptability Challenges

Figure 11.2 – A dedicated platform team structure

This team is responsible for the integration of new technologies, the optimization of existing processes, and the overall maintenance and enhancement of the platform. By focusing solely on the platform, the team can ensure that it not only meets the current needs of the organization but is also scalable and robust enough to support future requirements.

The team typically works in close coordination with other departments to gather requirements and feedback, ensuring that the platform evolves in alignment with the broader organizational goals. The team's activities might include automating testing processes, enhancing data security measures, implementing advanced analytics for better feedback loops, and ensuring the platform's compatibility with various tools and technologies.

Advantages

- **Specialized focus and expertise**: Having a team dedicated exclusively to the platform ensures a high level of specialized expertise, which can lead to more innovative solutions and advanced implementations.

- **Enhanced quality and performance**: The team's sole focus on the platform can lead to higher quality and performance, as continuous improvements and optimizations are systematically implemented.

- **Faster response times**: The dedicated nature of the team allows for quicker responses to technology-related issues, reducing downtime and improving overall operational efficiency.

- **Better alignment with business goals**: A dedicated team can more effectively align platform development with strategic business objectives, ensuring that the technology supports organizational needs at every level.

- **Increased scalability**: As the organization grows, the platform can be scaled to accommodate new demands without disrupting other processes thanks to the focused efforts of the dedicated platform team.

Disadvantages

- **Resource intensity**: Forming and maintaining a dedicated team requires significant resources, both in terms of personnel and budget. Smaller organizations may find this challenging.

- **Potential isolation**: There is a risk that the dedicated platform team could become isolated from the rest of the organization, leading to misalignments in understanding and objectives.

- **Dependency and bottlenecks**: Other departments might become overly dependent on the dedicated platform team, creating bottlenecks if the team becomes overloaded with requests or if their priorities do not align perfectly with immediate business needs.

- **Adaptability challenges**: The platform might evolve to become so specialized and tailored to current practices that it becomes less adaptable to new technologies or processes introduced into the market.

The dedicated platform team pattern is particularly advantageous for organizations with complex testing and operational needs that require ongoing attention and expertise. While it does involve significant investment and coordination, the benefits of having a robust, well-maintained, and continuously evolving platform can significantly outweigh the disadvantages. However, it's crucial for organizations considering this pattern to also implement strong communication channels between the dedicated platform team and other departments to avoid isolation and ensure that the platform remains aligned with the overall strategic goals of the organization.

Embedded teams

Embedded teams explore the dynamics and outcomes of running multiple teams in tandem, each being responsible for different segments of the project, yet all working towards a unified goal.

The embedded teams transformation implementation pattern, illustrated in *Figure 11.3*, represents a strategic approach designed to integrate specialized skills and capabilities directly within operational teams, promoting agility, enhanced collaboration, and the immediate application of expertise in the context of continuous improvement in testing, quality, security, and feedback processes. This pattern is increasingly favored in dynamic environments where quick adaptation and close collaboration across different functions are crucial.

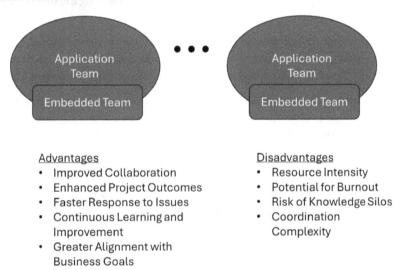

Figure 11.3 – The embedded teams transformation implementation pattern

Embedded teams consist of specialized professionals (such as quality assurance engineers, security experts, and feedback analysts) who are placed directly within functional project teams rather than being segregated into separate departments. These specialists work alongside developers, project managers, and operations staff throughout the entire project lifecycle. The primary goal of this arrangement is to foster a multidisciplinary approach to project challenges, ensuring that expertise in testing, quality, and security is not just an afterthought but a continuous and integral part of the development process.

This pattern is particularly effective in organizations adopting Agile methodologies, where the speed of development and iteration is rapid. Embedded specialists can immediately apply their skills to the tasks at hand, ensuring that best practices are followed from the outset of a project.

Advantages

- **Improved collaboration**: Having specialists embedded within teams fosters a culture of collaboration and knowledge sharing. This setup helps break down silos between departments, facilitating smoother communication and faster decision-making.

- **Enhanced project outcomes**: With experts in quality, security, and feedback integrated into the teams, projects benefit from high standards from the start, reducing the likelihood of costly reworks or security breaches post-deployment.

- **Faster responses to issues**: Immediate access to specialized knowledge allows teams to quickly identify and address potential issues before they escalate, enhancing the overall agility of the project management process.

- **Continuous learning and improvement**: Continuous interaction between specialists and other team members enhances the learning curve for the entire team. Non-specialists gain a better understanding of quality, security, and feedback considerations, which broadens their skill sets.

- **Greater alignment with business goals**: Since specialists are part of the core project teams, they are better positioned to align their strategic actions with the overarching business objectives, ensuring that all efforts contribute directly to organizational goals.

Disadvantages

- **Resource intensity**: Maintaining a high level of expertise within every project team can be resource-intensive. Smaller organizations might struggle with the availability of enough specialists to cover all teams.

- **Potential for burnout**: Specialists embedded in multiple teams or projects might face high demands on their time and expertise, leading to burnout and reduced effectiveness.

- **Risk of knowledge silos**: While the model aims to integrate knowledge, there's a risk that embedded specialists become gatekeepers of specialized knowledge rather than disseminators, leading to new kinds of silos within teams.

- **Coordination complexity**: Managing multiple embedded specialists across various projects can lead to coordination challenges, especially in larger organizations with numerous ongoing initiatives.

The embedded teams pattern offers a robust framework for integrating specialized expertise directly into project teams, fostering an environment of collaboration and immediate application of critical skills. This model is particularly effective in environments that value agility and rapid iteration, such as those using Agile methodologies. However, the approach requires careful management to avoid overextension of resources and ensure that the benefits of close integration do not translate into new challenges. Organizations considering this pattern should weigh the advantages of enhanced collaboration and project outcomes against the potential challenges of resource management and specialist burnout.

Outsourced teams

The third pattern is **outsourced teams**, which leverages external expertise and resources to accelerate project completion.

The outsourced teams transformation implementation pattern, shown in *Figure 11.4*, is an approach that involves delegating certain functions or projects to external organizations or teams rather than handling them in-house. This pattern is often adopted in scenarios where specific expertise is required, or where it may be more cost-effective to utilize external resources. In the context of enhancing organizational capabilities in continuous testing, quality, security, and feedback, outsourced teams can provide a strategic advantage by bringing in fresh insights and specialized skills that may not be available internally.

Advantages
- Access to Specialized Expertise
- Cost Efficiency
- Scalability
- Focus on Core Competencies
- Risk Mitigation

Disadvantages
- Control and Coordination Challenges
- Quality Risks
- Security Concerns
- Dependency
- Cultural and Organizational Misalignment

Figure 11.4 – The outsourced teams transformation implementation pattern

Outsourced teams are typically composed of third-party service providers or consultants who are contracted to manage and execute specific tasks within a project or operational function. This can include software testing, cybersecurity measures, quality assurance, and gathering and analyzing user feedback. The arrangement can be set up as a temporary project-based contract or as an ongoing service agreement, depending on the organization's needs.

Organizations opt for this pattern to leverage the specialized expertise of vendors who are leaders in their fields and to offload non-core activities, allowing internal teams to focus on strategic goals. The outsourced teams work in parallel with the internal teams, often under the supervision of project managers or coordinators who ensure that the services delivered align with the organization's standards and expectations.

Advantages

- **Access to specialized expertise**: Outsourcing allows organizations to access a pool of experts who can provide high-level skills and knowledge that are either lacking internally or are not economically feasible to develop

- **Cost efficiency**: It can be more cost-effective to hire outsourced teams, especially for short-term projects or highly specialized tasks, rather than employing full-time staff or investing in extensive training for existing employees.

- **Scalability**: Outsourced teams can be scaled up or down based on the current needs of the business, providing flexibility that is particularly valuable in environments with fluctuating demands.

- **Focus on core competencies**: By outsourcing non-core activities, an organization can focus its resources and energies on areas that offer the most strategic value and competitive advantage.

- **Risk mitigation**: Utilizing experienced outsourced teams can help mitigate risks, particularly in complex areas such as IT security, where the cost of errors can be very high.

Disadvantages

- **Control and coordination challenges**: Managing outsourced teams, especially those located in different time zones or with different cultural backgrounds, can lead to challenges in communication and coordination.

- **Quality risks**: Depending on the outsourced provider's commitment and capabilities, there may be risks associated with the quality of work, potentially leading to additional oversight and management costs.

- **Security concerns**: Sharing sensitive information or critical infrastructure components with an outsourced team can introduce security risks, especially if the outsourcing company does not adhere strictly to security protocols.

- **Dependency**: Over-reliance on outsourced teams can lead to dependency, which might be problematic if the outsourcing relationship ends abruptly or if the service provider's circumstances change.

- **Cultural and organizational misalignment**: Differences in work culture and organizational goals between the hiring company and the outsourced provider can lead to misunderstandings and misaligned outputs.

The outsourced teams pattern offers a practical approach for organizations looking to enhance their capabilities in specific areas without the need to develop those capabilities in-house. It provides access to specialized skills, potential cost savings, and flexibility. However, the success of this pattern largely depends on the choice of outsourcing partner, as well as the strength of the contractual agreements and management oversight put in place. Choosing the right pattern requires a careful assessment of the organization's needs, risks, and strategic objectives. When it is properly executed, outsourcing can be a valuable component of an organization's transformation strategy, particularly when it comes to scaling operations quickly and efficiently in areas such as continuous testing, quality, and security.

Hybrid dedicated/outsourced teams

The fourth and final pattern, **hybrid dedicated/outsourced teams**, combines elements of both dedicated internal teams and outsourced assistance to balance control, innovation, and scalability.

The hybrid dedicated/outsourced teams transformation implementation pattern, shown in *Figure 11.5*, merges the strengths of in-house expertise and external resources to foster a comprehensive approach to project management and execution. This pattern is increasingly popular in complex fields such as continuous testing, quality assurance, security, and customer feedback within organizations striving to enhance their capabilities while managing costs and maintaining control over critical processes.

Advantages	Disadvantages
• Balanced Control and Expertise	• Complex Coordination
• Cost Efficiency	• Cultural Differences
• Flexibility and Scalability:	• Quality Variability
• Enhanced Innovation	• Dependency on External Partners
• Risk Management	• Security and Confidentiality Risks

Figure 11.5 – The hybrid dedicated/outsourced teams transformation implementation pattern

Hybrid dedicated/outsourced teams consist of core in-house members who work closely with outsourced specialists to deliver specific project requirements. The internal team typically comprises employees who possess critical organizational knowledge and skills, and who are aligned with the company's culture and long-term goals. This team is complemented by external experts who provide specialized skills, innovative technologies, or cost efficiencies that might not be available internally.

This model allows organizations to maintain a strong foundation of dedicated staff while strategically utilizing external resources to enhance capacity or capabilities as needed. It effectively balances the control and familiarity of dedicated teams with the scalability and specialization of outsourced services.

Advantages

- **Balanced control and expertise**: This pattern provides a balance between maintaining control over key processes through the dedicated team and gaining access to specialized skills and new technologies through outsourced partners.

- **Cost efficiency**: By outsourcing non-core or highly specialized tasks, organizations can optimize costs without compromising the quality of the work or the strategic focus of the internal team.

- **Flexibility and scalability**: The hybrid model allows organizations to scale their operations up or down based on project demands without the long-term commitments associated with hiring full-time staff or the risks of fully outsourcing critical functions.

- **Enhanced innovation**: Access to external experts can bring fresh perspectives and innovative solutions that might not emerge from within an organization, fostering creativity and problem-solving.

- **Risk management**: Distributing tasks between dedicated and outsourced teams can help mitigate risks. If an outsourced provider fails to deliver, the internal team can step in to manage or rectify issues, ensuring continuity and stability.

Disadvantages

- **Complex coordination**: Managing a hybrid team requires robust coordination and communication strategies to ensure that both internal and external team members are aligned and working cohesively.

- **Cultural differences**: Integrating external resources with internal teams can lead to clashes in work culture, which may affect team dynamics and project outcomes.

- **Quality variability**: While outsourcing can bring in specialized skills, it can also lead to variability in quality depending on the external provider's standards and the oversight provided by the internal team.

- **Dependency on external partners**: There's a risk of becoming overly dependent on external providers, especially for specialized skills, which can be problematic if the outsourcing relationship is disrupted.

- **Security and confidentiality risks**: Sharing sensitive information and granting access to internal systems can expose the organization to increased security and confidentiality risks.

The hybrid dedicated/outsourced teams pattern offers a pragmatic solution for organizations aiming to enhance their operational capabilities while maintaining strategic control over key processes. It combines the depth of dedicated teams with the breadth of outsourcing to provide flexibility, cost efficiency, and access to specialized skills. However, the success of this pattern largely depends on effective integration strategies, strong management practices, and careful selection of outsourcing partners.

In summary, while some implementation patterns are preferable to others, the importance of selecting the right pattern cannot be overstated. It ensures that the organizational transformation not only takes place efficiently but also aligns with and supports the broader goals of the organization effectively.

Patterns to avoid during implementation

In the quest to enhance organizational capabilities, especially in areas such as continuous testing, quality, security, and feedback, certain implementation patterns, as shown in *Figure 11.6*, have consistently proven to be less effective or even detrimental. Identifying and avoiding these unfavorable implementation patterns is as critical as adopting effective ones is.

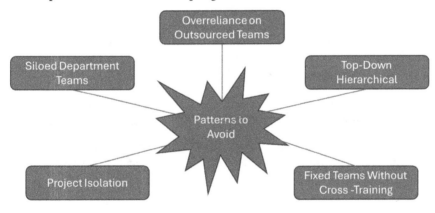

Figure 11.6 – Patterns to avoid

The following are some common patterns to avoid, along with detailed explanations of their drawbacks:

- **Siloed department teams**: Siloed department teams operate independently with minimal interaction or communication with other departments. Each department focuses solely on its specific functions without integrating its activities with the broader organizational goals. The drawbacks are as follows:

  - **Lack of communication**: Silos create barriers to communication, leading to misaligned objectives across departments. This can impede the flow of information that is necessary for effective decision-making.

- **Inefficiency**: Duplicate efforts and reinvention of solutions that other parts of the organization may have already developed can occur, leading to inefficiency and wasted resources.

  - **Delayed responses**: When issues arise that require cross-departmental cooperation, the response can be slow and cumbersome, impacting the overall agility of the organization.

- **Over-reliance on outsourced teams**: This pattern occurs when organizations depend heavily on external entities for critical functions and processes without maintaining sufficient in-house expertise. The drawbacks are as follows:

  - **Loss of control**: Over-reliance on outsourced teams can lead to a loss of control over critical business processes and potentially diminish the organization's ability to make agile decisions.

  - **Knowledge drain**: The organization risks losing internal competency and institutional knowledge. This can be detrimental in the long term, especially when contracts with outsourcing providers end.

  - **Security risks**: Entrusting sensitive data and processes to third-party providers increases the risk of data breaches and security lapses.

- **Top-down hierarchical implementation**: This pattern is characterized by decision-making and implementation processes that flow downward from the top of the organization without adequate input or engagement from lower-level staff. The drawbacks are as follows:

  - **Resistance to change**: Changes imposed from the top without involving those who execute them often meet with resistance, reducing the effectiveness of implementation.

  - **Lack of innovation**: Lower-level employees are often closer to the day-to-day operations and could have innovative ideas that are overlooked because there isn't a mechanism for upward communication. There can be a fear of proposing new ideas in top-down organizational structures.

  - **Slow adaptation**: The rigidity of top-down approaches can hinder the organization's ability to adapt quickly to changing market conditions or internal challenges.

- **Fixed teams with no cross-training**: Teams in this pattern work in fixed roles with specific tasks, with little to no effort made to cross-train members on other functions or roles within the team. The drawbacks are as follows:

  - **Vulnerability**: The absence of cross-training can lead to vulnerabilities, especially if key team members are unavailable, as others cannot effectively cover their roles.

  - **Reduced team flexibility**: Lack of versatility in team skills can limit the team's ability to adapt to new challenges or changes in project scope.

  - **Stagnation in skill development**: Employees might feel their growth and learning are stifled if they are confined to a single role or set of tasks, leading to decreased job satisfaction and higher turnover.

- **Project isolation**: In this pattern, projects are managed in isolation from one another without sharing learnings, resources, or synergies between them. The drawbacks are as follows:

  - **Reinvention of the wheel**: Each project team may end up solving problems that have already been addressed by others, wasting time and resources.

  - **Missed opportunities for synergy**: Isolated projects fail to leverage the collective knowledge and resources of the organization, missing out on potential efficiencies. This can lead to bottlenecks, limiting the capacity and flow of the organization.

  - **Inconsistency in standards and quality**: Without a unified approach, different projects might adopt varying standards and practices, leading to inconsistencies in quality and performance across the organization.

Avoiding these unfavorable implementation patterns is essential for building a resilient and adaptive organization. By fostering open communication, maintaining a balanced approach to outsourcing, encouraging cross-functional teams, and promoting integration across projects, organizations can enhance their operational effectiveness and adapt more swiftly to new challenges and opportunities. Each of these patterns highlights the importance of strategic decision-making in organizational transformations, emphasizing the need to carefully design and manage the processes that determine how teams and projects are structured and executed.

Selecting an implementation pattern

Selecting the right implementation pattern is crucial for organizations aiming to enhance their operational capabilities, particularly in the realms of continuous testing, quality, security, and feedback processes. The process, shown in *Figure 11.7*, involves careful consideration of organizational needs, existing capabilities, and strategic goals.

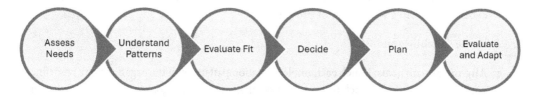

Figure 11.7 – Selecting an implementation pattern

The following is a detailed, recommended process to guide organizations in choosing an effective implementation pattern that aligns with their specific circumstances and objectives:

1. **Assess organizational needs and capabilities**:

    *   **Objective assessment**: Begin with a thorough assessment of the current state of the organization. This includes evaluating existing processes, resources, and technologies. Identify **Strengths, Weaknesses, Opportunities, and Threats** (**SWOT** analysis) to understand where the organization stands and what it needs to improve.

    *   **Identify goals**: Clearly define what the organization aims to achieve with the transformation. Goals should be **specific, measurable, achievable, relevant, and time-bound** (**SMART**). This might include speeding up release cycles, improving product quality, enhancing security measures, or refining feedback mechanisms.

    *   **Skill and resource inventory**: Take inventory of the available skills and resources. This helps determine whether the organization has the necessary capabilities in-house or whether it needs to look externally to fill gaps.

2. **Understand different implementation patterns**:

    *   **Research patterns**: Gather detailed information on various implementation patterns that have been successful in similar organizations or industries. This research should include case studies, best practices, and even lessons learned from less successful implementations.

    *   **Consult stakeholders**: Engage with various stakeholders, including IT staff, project managers, department heads, and potentially even customers. Their insights can provide valuable perspectives on what might work best for the organization.

    *   **Expert consultations**: It may be beneficial to consult with external experts or consultants who specialize in organizational transformation. They can offer unbiased advice and share experiences from a wide range of projects.

3. **Evaluate suitability and fit**:

    *   **Alignment with goals**: Match each implementation pattern with the organization's predefined goals. Evaluate how well each pattern can potentially meet these goals considering the current and future states of the organization.

    *   **Feasibility study**: Conduct a feasibility study for each potential pattern. Analyze the logistical, financial, and technical aspects of implementing these patterns. This study should also consider the potential risks and mitigation strategies for each pattern.

    *   **Pilot testing**: If possible, conduct pilot tests for the most promising patterns. Pilots can provide practical insights into how a pattern might work within the organization and reveal any adjustments needed before a full-scale rollout.

4. **Decision-making**:

   - **Collective decision-making**: Utilize a decision-making body or committee that includes representatives from all relevant stakeholders. This group can review the findings from the feasibility studies and pilot tests to make informed decisions.

   - **Use decision matrices**: Implement decision-support tools such as decision matrices that weigh various factors such as cost, impact, risk, and alignment with strategic goals. These tools can help in making objective decisions.

   - **Feedback loops**: Establish feedback mechanisms throughout the decision-making process to ensure that all concerns and insights are considered and addressed.

5. **Plan implementation**:

   - **Detailed planning**: Once a pattern has been selected, develop a detailed implementation plan. This plan should outline all necessary steps, timelines, resource allocations, and responsibilities.

   - **Change management**: Develop a change management strategy to help manage the transition. This should include communication plans, training programs, and support structures to ensure smooth adoption by all affected parties.

   - **Risk management**: Identify potential risks associated with the implementation and develop strategies to mitigate these risks. This could involve contingency planning and establishing escalation paths for dealing with issues as they arise.

6. **Continuous monitoring and adaptation**:

   - **Monitor progress**: Regularly monitor the implementation's progress against the set goals and timelines. Use **Key Performance Indicators** (**KPIs**) to measure success and identify areas needing improvement.

   - **Iterative improvement**: Be prepared to make iterative improvements based on feedback and the outcomes of ongoing evaluation. This adaptive approach helps refine the implementation process and better align it with organizational needs.

Selecting the right implementation pattern requires a systematic and thorough approach that considers an organization's unique context and objectives. By carefully assessing needs, as well as understanding available patterns, evaluating their fit, and planning meticulously for their implementation, organizations can significantly enhance their chances of success in transforming their operations. This structured selection process not only aids in choosing the most appropriate pattern but also ensures its effective integration and sustainability within the organization.

Summary

Chapter 11 provided a comprehensive exploration of implementation patterns that are instrumental for organizations looking to advance their capabilities in continuous testing, quality, security, and feedback. The chapter began by defining what an implementation pattern is—it's essentially a blueprint that, when followed, enhances the likelihood of achieving project objectives effectively and efficiently. The discussion emphasized the critical role these patterns play in ensuring successful project execution and set the stage for an in-depth review of various effective patterns and those that should be avoided.

Four primary successful implementation patterns were highlighted: dedicated platform team, embedded teams, outsourced teams, and hybrid dedicated/outsourced teams. Each pattern was described in detail, providing insights into their unique advantages and how they contribute to the success of organizational projects. For instance, the chapter explained that the dedicated platform team approach focuses on the centralized handling of platform development, ensuring high quality and performance, while embedded teams promote agility and close collaboration across different functions within projects. Outsourced teams bring in external expertise and flexibility, and hybrid teams combine the strengths of both in-house and outsourced resources to optimize project execution.

The chapter also discussed implementation patterns that organizations should avoid, such as siloed department teams, over-reliance on outsourced teams, top-down hierarchical implementation, fixed teams with no cross-training, and project isolation. Each of these patterns was analyzed for their drawbacks, such as inefficiencies, increased risks, and the challenges they present in dynamic business environments. The focus was on understanding why these patterns hinder progress and how they can negatively impact the organization's ability to adapt and innovate.

The chapter provided practical guidance on how to select the most appropriate implementation pattern based on an organization's specific needs, goals, and resources. This guidance is vital for decision-makers to ensure that the chosen pattern aligns well with strategic objectives and operational demands. The next chapter will build on this foundation by exploring how to effectively measure progress and outcomes, ensuring that organizations can monitor the success of the implementation patterns and make necessary adjustments to achieve their transformation goals.

12

Measuring Progress and Outcomes

This chapter focuses on methods and frameworks that are important for measuring progress and outcomes as organizations implement and improve their continuous testing, quality, security, and feedback capabilities. As businesses and projects become more complex, it's crucial to be able to measure how well things are going and to predict future results. This chapter explains how to create, choose, and use measurements that accurately show progress and offer useful insights for management and continuous improvement.

The first section provides an overview of the different kinds of measurements for continuous testing, quality, security, and feedback. Then, we will discuss how to select the right measurements for both outcomes and progress. Then, we will move on to implementing these metrics and look at how to set up metrics and dashboards that do more than just inform – they motivate action.

We will wrap up the chapter by discussing how to keep these measurements useful over time, offering strategies to update and maintain them so that they keep supporting your organization as it grows and changes. This chapter will equip you with the skills to not only select effective measures but also to apply and maintain them, in ways that continuously add value to your work.

In this chapter, we'll cover the following main topics:

- Measures of progress and outcomes
- Selecting measures
- Practices to design metrics and dashboards
- Sustaining measures of progress and outcomes

Let's get started!

Measures of progress and outcomes

In this section, we will understand what makes a measurement effective and how to align it with your organization's goals. By exploring different measurements, you'll get a solid base for later discussions on how to choose and use these metrics.

Organizations that undergo transformation processes for continuous testing, quality, security, and feedback need to keep track of two key types of measures – *progress* and *outcomes*. These measures are crucial for understanding how well an organization is adapting and improving its processes, and for ensuring that these changes contribute positively to its overall goals.

Figure 12.1 illustrates the two types of metrics, in chart form.

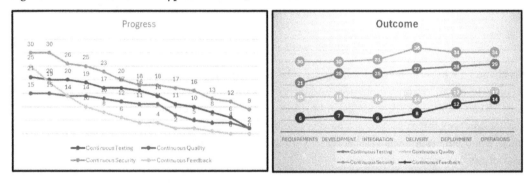

Figure 12.1 – Progress and outcome metric charts

Progress metrics are typically represented in the form of burn-down charts, in which a declining trend is desirable because they indicate how many project units are remaining. Outcome metrics are typically represented as a trend line, in which an increasing trend is desirable as performance increases.

The next section will explain the importance of measures of progress and outcomes.

Why measures of progress and outcomes are important

There are several reasons why measures of progress and outcomes are important, including strategic alignment and decision making, performance monitoring, demonstrating value, and to help create a learning organization.

- **Strategic alignment and decision making**: Knowing how far along transformation projects and initiatives are helps companies make better decisions about where to focus their resources. It ensures that every effort contributes toward the broader goals of an organization.

- **Performance monitoring**: Organizations need to check regularly whether they are on the right path to meeting their goals. Measures of progress help by showing how much has been done and how much is left to do. This is especially important in environments where updates and improvements are constant, such as in software development or cybersecurity.

- **Demonstrating value**: Outcome measures inform a company whether any changes made have had the desired effect. This is important not just for internal records but also to show external stakeholders, such as customers or investors, that the company is improving and worth investing in.

- **Organizational learning**: Both types of measures provide valuable feedback. This feedback helps companies learn from their successes and mistakes, leading to better planning and more effective management of risks.

Next, we will explain how to link measures of performance and progress to the capability maturity of an organization.

Linking measures to capability maturity

The concept of capability maturity in an organization, as explained in *Chapter 4*, relates to its ability to consistently deliver expected results. Progress and outcome measures are integral to this because they help an organization understand its current level of maturity, as described in *Chapter 4*, and identify areas for improvement. For instance, as an organization matures, it should see more predictable and positive outcomes from its processes, which should be evident in the measures it tracks. Tracking these measures over time can show how the organization's capability maturity evolves. *Figure 12.2* illustrates important factors and stakeholders for progress and outcome measures.

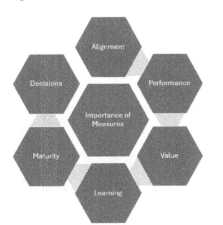

Stakeholders
- Business Managers
- Technical Managers
- Developers
- QA Engineers
- Security Staff
- DevOps engineers
- SREs

Figure 12.2 – The importance of measures to stakeholders

The following explains why these measures are important for various stakeholders:

- **Business managers**: These managers look at outcome measures to see whether the changes positively affect a business's bottom line. They use progress measures to ensure that strategic projects are on schedule.

- **Technical managers and developers**: They rely on progress measures to plan and adjust their workloads and timelines. Outcome measures help them see whether their technical solutions are effective in solving the problems they are meant to address.

- **Quality assurance (QA) engineers**: QA teams use these measures to track and improve the testing processes. They need to know that the quality of the product meets the set standards, which is directly tied to outcome measures.

- **Security operations staff**: For these teams, outcome measures might include metrics such as a reduced number of security incidents. Progress measures help them gauge the implementation of security protocols and training.

- **DevOps and SRE engineers**: These engineers use both types of measures to optimize the deployment and operation of software. Progress measures help them monitor the continuous delivery pipeline, while outcome measures evaluate the stability and performance of the software in production.

In summary, measures of progress and outcomes provide a structured way to track how well an organization implements changes and achieves results. They offer critical insights that help various stakeholders, from business managers to technical teams, make informed decisions that drive the organization forward. Understanding and utilizing these measures effectively can lead to better strategic alignment, improved performance, and a mature capability to deliver consistent results.

Examples of outcome metrics

Outcome metrics are essential tools in software development, delivery, and operations. They help teams track how well they are doing at various stages of building and maintaining software. These metrics focus on areas such as testing, quality, security, and feedback to ensure that every phase, from planning to operations, meets set standards and improves over time.

Common outcome metrics across all stages

The **DevOps Research and Assessment** group created a set of metrics known as DORA which are a set of four key performance indicators that have become industry standards for measuring the effectiveness of software development and operations teams.

These metrics, recommended by DORA, are broadly applicable and provide insights into the overall health and efficiency of the development process:

- **Deployment frequency (DORA)**: Counts how often software is deployed to production.

- **Lead time for changes (DORA)**: Measures the time from code commit to deployment.

- **Change failure rate (DORA)**: Calculates the percentage of deployments that fail.

- **Mean time to recover (MTTR) (DORA)**: Gauges the average time taken to recover from a failure.

- **Availability**: Tracks the percentage of time an application is available without issues.

Figure 12.3 shows that the outcome metrics apply to every stage in the value stream.

	Requirement	Development	Integration	Delivery	Deployment	Operations
Deployment Frequency	x	x	x	x	x	x
Lead Time	x	x	x	x	x	x
Change Failure Rate	x	x	x	x	x	x
Mean Time to Recover	x	x	x	x	x	x
Availability	x	x	x	x	x	x

Figure 12.3 – Common outcome metrics across all stages

The importance of DORA Metrics

DORA metrics are important because they provide an objective measurement of outcomes, including measures of DevOps performance, a benchmark for improvements, and alignment with business goals, as follows:

- **Objective measurement of DevOps Performance**: DORA metrics provide quantitative data that organizations can use to assess the performance of their DevOps practices objectively. These metrics have been validated across various organizations and are shown to correlate with higher software delivery performance and organizational performance.

- **Benchmarks for improvement**: By establishing benchmarks based on these metrics, organizations can gauge their current performance and identify areas where they need to improve. This can lead to more targeted investments in technology, processes, and training, ultimately leading to more efficient and effective development and operational practices.

- **Alignment with business goals**: High performance on DORA metrics generally correlates with higher organizational efficiency, better product quality, faster time to market, and improved customer satisfaction. These aspects are crucial for achieving broader business goals, such as growth, profitability, and competitive advantage.

The utility of DORA metrics across software value stream stages

DORA metrics, with some adaptations, can be applied across the entire end-to-end value stream and each stage in the value stream, as follows:

- **Requirements stage**: While DORA metrics are traditionally more aligned with the deployment stage of the software development life cycle, they indirectly influence the requirements stage by ensuring that the processes that follow are streamlined and efficient, thereby enabling quicker and more reliable requirement changes and deployments.

- **Development and continuous integration stages**: As indicated previously, DORA metrics are traditionally more aligned with the deployment stage of the software development life cycle. With some logical revisions, DORA metrics can be used to measure the effectiveness of the development and continuous integration stages, such as the following:

  - **Lead time for changes** reflects the efficiency of the development process, including how quickly code changes can be integrated, tested, and ready for release.

  - **Deployment frequency** during these stages can indicate the health of the CI/CD pipeline, showing how mature continuous testing practices are and how often integrations and potential releases occur, even if they are to staging environments rather than production.

- **Continuous delivery and deployment stages**:

  - **Deployment frequency** directly measures how often deployments occur, which is crucial in these stages.

  - **Change failure rate** provides insights into the maturity and quality of SDLC practices and deployment processes, indicating how often deployments lead to operational disruptions that need quick fixes.

  - MTTR is crucial here, as it shows a team's capability to quickly resolve issues that occur after code is deployed to production or staging environments.

- **Operations stage**:

  - MTTR shines in the operations stage by measuring operational resilience and the ability to quickly restore service after incidents.

  - **Change failure rate** also helps assess the stability of the production environment, as frequent failures can indicate systemic problems that might originate from earlier stages of the value stream.

By providing a clear picture of these key aspects, DORA metrics help organizations not only measure outcomes but also understand the flow of value through the entire software development life cycle. They enable teams to iterate on and refine their practices, promoting a cycle of continuous improvement that is aligned with both technical and business outcomes.

Continuous testing outcome metrics

Continuous testing ensures that every part of the software is tested at every stage of its development to catch and fix issues early. Examples of these metrics are illustrated in *Figure 12.4* and described as follows:

	Requirement	Development	Integration	Delivery	Deployment	Operations
Test Case Creation Speed	x	x				x
Code Coverage		x				
Build Pass Rate			x			
Test Automation Rate		x	x	x	x	x
Deployment Health					x	x
Operations Test Speed						x

Figure 12.4 – Continuous testing outcome metrics by stage

- **Requirements stage**: *Test case creation speed* – measures how quickly test cases are created for new requirements.

- **Development stage**: *Code coverage* – calculates the percentage of the code base tested by automated tests.

- **CI stage**: *Build pass rate* – tracks the percentage of builds that pass all tests.

- **Continuous delivery stage**: *Test automation rate* – measures the percentage of tests that are automated.

- **Continuous deployment stage**: *Deployment health checks* – assesses the success of deployments based on automated post-deployment health checks.

- **Continuous operations stage**: *Post-deployment testing speed* – measures how quickly tests are conducted and results returned after deployment.

As you can see from *Figure 12.4*, some of the continuous testing metrics are best applied at specific stages of the software value stream. In the next section, we will see that this is also true for continuous quality metrics.

Continuous quality outcome metrics

Continuous quality metrics, illustrated in *Figure 12.5*, focus on maintaining high standards throughout the life cycle of the software.

	Requirement	Development	Integration	Delivery	Deployment	Operations
Requirement Clarity Index	x					
Defect Discover Rate		x				
Integration Error Rate			x			
Release Quality Score				x		
User Experience Scores					x	
System Down Time						x

Figure 12.5 – Continuous quality outcome metrics by stage

- **Requirements stage**: *Requirements clarity index* – Rates how clear and understandable the project requirements are.

- **Development stage**: *Bug discovery rate* – tracks the rate at which bugs are found during development.

- **Continuous integration stage**: *Integration error rate* – measures the frequency of errors during the integration phase.

- **Continuous delivery stage**: *Release quality score* – evaluates the quality of software at the release based on predefined criteria.

- **Continuous deployment stage**: *User experience scores* – tracks activities affecting user experience discovered after deployment.

- **Continuous operations stage**: *System downtime* – measures the total time the system is non-operational due to quality issues.

As you can see, it is a good practice to have a continuous quality metric for each stage of the value stream. The same is true for continuous security metrics also.

Continuous security outcome metrics

Security metrics, illustrated in *Figure 12.6*, ensure that software is protected against threats at every stage.

	Requirement	Development	Integration	Delivery	Deployment	Operations
Security Requirements Completion	x					
Vulnerabilities per Line of Code		x				
Security Scan Pass Rate			x			
Critical Security Issue Rate				x		
Security Incident Response Time					x	
Security Compliance Rate						x

Figure 12.6 – Continuous security outcome metrics by stage

- **Requirements stage**: *Security requirements completion* – tracks how completely security requirements are met in the planning phase.

- **Development stage**: *Vulnerabilities per line of code* – measures the number of security vulnerabilities relative to the size of the code base.

- **CI stage**: *Security scan pass rate* – the percentage of passes in routine security scans during integration.

- **Continuous delivery stage**: *Critical security issue rate* – tracks the rate of critical security issues found before delivery.

- **Continuous deployment stage**: *Security incident response time* – measures how quickly security incidents are addressed after deployment.

- **Continuous operations stage**: *Ongoing security compliance rate* – measures compliance with ongoing security standards and regulations.

As explained in this section, continuous security metrics are recommended for each stage in the value stream. The next section shows that this is also true for continuous feedback metrics.

Continuous feedback outcome metrics

Feedback metrics, as shown in *Figure 12.7*, evaluate how effectively an organization gathers and implements feedback from users throughout the development process.

	Requirement	Development	Integration	Delivery	Deployment	Operations
Feedback Integration Effectiveness	x					
Developer Response Time to Feedback		x				
Feedback Resolution Rate			x			
Feedback Impact Score				x		
Customer Satisfaction Post-Deployment					x	
Long-term Feedback Trends						x

Figure 12.7 – Continuous feedback outcome metrics by stage

- **Requirements stage**: *Feedback integration effectiveness* – measures how effectively user feedback is integrated into requirements.

- **Development stage**: *Developer response time to feedback* – tracks how quickly developers address feedback during the development.

- **Continuous integration stage**: *Feedback resolution rate* – measures the rate at which feedback issues are resolved during integration.

- **Continuous delivery stage**: *Feedback impact score* – evaluates the impact of implemented feedback on software quality.

- **Continuous deployment stage**: *Customer satisfaction post-deployment* – measures customer satisfaction levels after deploying new features or fixes.

- **Continuous operations stage**: *Long-term feedback trends* – analyzes trends in user feedback over the operation phase to guide future improvements.

Overall, outcome metrics help teams monitor and improve the efficiency, security, and quality of their software from start to finish, ensuring that each stage of the value stream is measured and can be optimized to meet both user needs and business goals.

Examples of progress metrics

Progress metrics are critical in tracking the advancement of transformation and software development projects. They help teams monitor the efficiency and effectiveness of their work at different stages of the software life cycle. These metrics are specifically designed to measure how projects progress against planned schedules, goals, and benchmarks in areas such as testing, quality, security, and feedback.

Common progress metrics across all stages

Progress metrics play a pivotal role in software development, providing teams with real-time insights into the progress and efficiency of their projects. These metrics help track advancement through various stages of the software life cycle, including testing, quality, security, and feedback, ensuring that projects meet deadlines and benchmarks.

These metrics, including key flow metrics (as shown in *Figure 12.8*), help teams monitor overall project health and progress, ensuring that timelines and productivity goals are met.

	Requirement	Development	Integration	Delivery	Deployment	Operations
Sprint Burndown Chart	x	x	x	x	x	x
Release Burndown	x	x	x	x	x	x
Cumulative Flow Diagram (Flow Metric)	x	x	x	x	x	x
Velocity	x	x	x	x	x	x
Work in progress (WIP) Limits (Flow Metric)	x	x	x	x	x	x
Cycle Time (Flow Metric)	x	x	x	x	x	x

Figure 12.8 – Progress metrics relevant to all stages of the value stream

The following are examples of useful progress metrics and visualization artifacts:

- **Sprint burndown chart**: Shows the remaining work in the sprint versus time, indicating whether the team is on track.

- **Release burndown**: Measures the amount of remaining work against the timeline for the release.

- **Cumulative flow diagram (flow metric)**: Visualizes the quantity of work in different stages of development, helping to identify bottlenecks and understand work balance and flow through a system.

- **Velocity**: Assesses the amount of work a team completes during a sprint, which is useful for future sprint planning.

- **Work in progress (WIP) limits (flow metric)**: Sets the maximum amount of tasks that can be in progress at any one time, helping teams manage workload and reduce bottlenecks.

- **Cycle time (flow metric)**: Measures the time it takes for work to move from start to finish, providing insights into process efficiency.

Flow metrics are essential indicators used to measure the movement of work items through various stages of the software development life cycle. These metrics focus on the efficiency, effectiveness, and predictability of software delivery processes.

The importance of flow metrics

Flow metrics help organizations understand how work progresses from initial concept to delivery in a tangible and actionable way:

- **Enhancing visibility**: Flow metrics provide clear visibility into the current state of work processes. By tracking the progression of work items, teams can identify where bottlenecks or delays occur and understand their root causes.

- **Improving efficiency**: With insights from flow metrics, organizations can streamline processes, remove inefficiencies, and optimize resource allocation. This can lead to quicker development cycles and more efficient use of time and resources.

- **Boosting predictability**: Flow metrics allow teams to predict future performance based on historical data. This predictability helps in better planning and forecasting, reducing uncertainties in delivery timelines.

- **Promoting continuous improvement**: Regular measurement and analysis of flow metrics encourage a culture of continuous improvement. Teams can iteratively refine their processes based on quantifiable data, striving for leaner and more efficient workflows.

- **Alignment with business objectives**: By improving efficiency and predictability, flow metrics help ensure that software delivery is aligned with broader business goals, such as increasing customer satisfaction, reducing time to market, and improving product quality.

The utility of flow metrics across software value stream stages

The following details the recommended applications of flow metrics for each stage in the software value stream:

- **The requirements stage**:

  - **Work item age**: Measures how long a requirement has been under discussion before being approved. This metric helps identify delays in the early stages of the value stream.

- **Flow efficiency**: The ratio of active work time to the total elapsed time. Low efficiency at this stage might indicate too much wait time or indecision in defining requirements.

- **The development stage**:

  - **WIP**: Limits the number of tasks being worked on simultaneously to reduce context switching and focus on completing tasks quicker.

  - **Cycle time**: Measures the time it takes for work to move from the start of development to completion. Monitoring cycle time helps in identifying slowdowns and improving developer productivity.

- **The CI/CD stage**:

  - **Throughput**: Tracks the number of work items passing through the CI/CD pipeline during a given period. High throughput indicates a healthy CI/CD pipeline.

  - **Lead time**: The time from work item initiation to delivery. It includes cycle time and all forms of delay. Reducing lead time at this stage is critical for faster releases.

- **The deployment and operations stage**:

  - **Release frequency**: Measures how often new releases are deployed to production. Frequent releases can indicate a smooth flow from development to deployment.

  - **MTTR**: Although traditionally a DORA metric, MTTR in the context of flow metrics helps measure the speed at which a team can recover from failures in production, reflecting the operational agility.

In all stages, the application of flow metrics provides a clear, data-driven view of how tasks are progressing, where issues are occurring, and what improvements are necessary. This level of insight is invaluable for any organization undergoing transformation and aiming to enhance its software delivery and operational capabilities continuously. By measuring and optimizing each stage of the software value stream, organizations can achieve a smoother, faster, and more reliable flow of value to their customers.

Continuous testing progress metrics

Progress metrics in continuous testing, as shown in *Figure 12.9*, track the development and execution of tests to ensure timely quality assessments.

	Requirement	Development	Integration	Delivery	Deployment	Operations
Test Case Preparation Progress	x					
Test Execution		x				
Build Integration Progress			x			
Test Automation Progress				x		
Deployment Progress Tracking					x	
Monitoring System Implementation Progress						x

Figure 12.9 – Continuous testing progress metrics

- **Requirements stage**: *Test case preparation progress* – tracks the percentage of test cases prepared against the total planned.

- **Development stage**: *Test execution progress* – measures the percentage of tests run against the total tests planned.

- **Continuous integration stage**: *Build integration progress* – tracks the number of successful integrations against planned integrations.

- **Continuous delivery stage**: *Test automation progress* – monitors the percentage of tests automated against the target.

- **Continuous deployment stage**: *Deployment progress tracking* – measures the frequency and success rate of deployments against the schedule.

- **Continuous operations stage**: *Monitoring system implementation progress* – tracks the implementation and fine-tuning of system monitoring tools.

Here, we looked at the recommended progress metrics for continuous testing. In the next section, we will do the same for continuous quality progress metrics.

Continuous quality progress metrics

Continuous quality progress metrics, as shown in *Figure 12.10*, focus on tracking the progress of efforts to ensure and enhance a product's quality throughout its life cycle.

	Requirement	Development	Integration	Delivery	Deployment	Operations
Requirements Review	x					
Code Review Coverage		x				
Code Quality Improvement			x			
Pre-release Testing Completion				x		
Post-deployment Quality Checks					x	
Quality Metrics Monitoring						x

Figure 12.10 – Continuous quality progress metrics

- **Requirements stage**: *Requirements review completion* – the percentage of project requirements reviewed and approved.

- **Development stage**: *Code review coverage* – tracks the percentage of new code that has been reviewed.

- **CI stage**: *Code quality improvement* – measures improvements in code quality metrics over time.

- **Continuous delivery stage**: *Pre-release testing completion* – the percentage of planned testing completed before release.

- **Continuous deployment stage**: *Post-deployment quality checks* – tracks the completion of quality checks after each deployment.

- **Continuous operations stage**: *Quality metrics monitoring* – the percentage of operational quality metrics being monitored as planned.

Here, we looked at the metrics to track continuous quality progress. Next, we will discuss the metrics to demonstrate continuous security progress.

Continuous security progress metrics

Continuous security progress metrics, as shown in *Figure 12.11*, focus on monitoring the progress of integrating and improving security measures throughout the development process.

	Requirement	Development	Integration	Delivery	Deployment	Operations
Security Plan Completion	x					
Security Integration Progress		x				
Security Test Execution			x			
Security Audit Completion				x		
Security Deployment Checks					x	
Security Update Implementation						x

Figure 12.11 – Continuous security progress metrics

- **Requirements stage**: *Security plan completion* – measures how much of the security planning has been completed.

- **Development stage**: *Security integration progress* – tracks the integration of security features into the product.

- **Continuous integration stage**: *Security test execution* – the percentage of planned security tests that have been executed.

- **Continuous delivery stage**: *Security audit completion* – measures the completion of security audits before delivery.

- **Continuous deployment stage**: *Security deployment checks* – tracks the completion of security checks during deployment.

- **Continuous operations stage**: *Security update implementation* – measures the implementation rate of security updates and patches.

Here, we looked at the metrics to demonstrate the progress of continuous security. The next section will suggest metrics to demonstrate the progress of continuous feedback.

Continuous feedback progress metrics

Continuous feedback progress metrics, as shown in *Figure 12.12*, assess how efficiently feedback is collected and acted upon throughout the development process.

	Requirement	Development	Integration	Delivery	Deployment	Operations
Feedback Collection Completion	x					
Feedback Implementation Progress		x				
Feedback Integration Rate			x			
Feedback Review Completion				x		
Customer Feedback Analysis					x	
Ongoing Feedback Monitoring						x

Figure 12.12 – Continuous feedback progress metrics

- **Requirements stage**: *Feedback collection completion* – tracks the percentage of planned feedback collection activities completed.

- **Development stage**: *Feedback implementation progress* – measures the progress of implementing feedback into the development process.

- **CI stage**: *Feedback integration rate* – tracks the rate at which feedback is integrated during integration testing.

- **Continuous delivery stage**: *Feedback review completion* – the percentage of feedback review tasks completed before delivery.

- **Continuous deployment stage**: *Customer feedback analysis* – measures the completion of post-deployment customer feedback analysis.

- **Continuous operations stage**: *Ongoing feedback monitoring* – tracks the ongoing monitoring and analysis of user feedback.

These progress metrics provide teams with the necessary tools to ensure that they are moving toward their project goals efficiently, allowing for timely adjustments in strategy and execution to meet the desired outcomes.

Selecting measures

This is where we apply what we've learned; you'll see how to tell apart measurements that only provide information from those that can truly transform your processes. We'll focus on making sure the measurements you pick are relevant, reliable, and able to drive significant changes.

When selecting and prioritizing outcomes or progress metrics for a project or organization, it's crucial to follow a systematic process to ensure the metrics chosen effectively measure success and align with business objectives. *Figure 12.13* shows a recommended approach to do so.

Figure 12.13 – Selecting outcome and progress metrics

The following is the recommended process to select both outcome and progress metrics.

- **Define business and project objectives**: Start by clearly defining the business objectives. Understanding what an organization aims to achieve with its projects or processes is critical. Whether it's improving customer satisfaction, increasing software deployment frequency, enhancing security measures, or reducing operational costs, having a clear set of goals will guide the selection of relevant metrics.

- **Identify metrics**: Once business or project objectives are defined, identify the key metrics that directly reflect success toward these objectives. Metrics should be actionable, meaning they can guide decision-making and influence outcomes. For example, if the objective is to enhance customer satisfaction, a possible KPI could be the **Net Promoter Score** (**NPS**). The metrics described earlier in this chapter provide suggestions of both outcome and progress metrics that you can choose from.

- **Ensure metrics are measurable and achievable**: Select metrics that are not only aligned with business objectives but are also measurable and achievable. Each metric should have a clear method of measurement and be grounded in data that can be consistently and accurately collected. Avoid vague metrics that are open to interpretation, as they can lead to inconsistent results and decisions.

- **Prioritize metrics based on impact and relevance**: Not all metrics are created equal. Some will have a greater impact on business outcomes than others. Prioritize metrics based on their potential impact on business goals and their relevance to current projects or strategies. Metrics that offer insights into customer behavior, product performance, and operational efficiency often take precedence, as they directly correlate with business success.

- **Consider stakeholder input**: Engage with stakeholders across different departments of an organization to get their input on which metrics they believe are most important. This not only ensures a holistic view but also facilitates buy-in across the organization. Stakeholders can provide insights into the practical aspects of metric tracking and implications that might not be obvious at the strategic level.

- **Review and refine regularly**: Outcome metrics should not be static. As business goals shift, technologies evolve, and markets change, the relevance of different metrics will vary. Regularly review metrics to ensure they remain aligned with business objectives and make adjustments as necessary. This might mean refining the way metrics are measured or replacing less relevant metrics with ones that better match the current business landscape.

By following these steps, organizations can effectively select and prioritize outcome metrics that not only provide a clear indication of performance but also drive meaningful improvements aligned with strategic goals. The set of metrics should be relevant and balanced.

Leadership and teams for selecting outcome and progress metrics

Selecting both outcome and progress metrics for projects involves strategic decision-making and should include input from various roles within an organization. It's important to have a balanced team of leaders and team members to ensure that the metrics chosen align with business goals, project objectives, and practical implementation considerations. Here's a guideline on who should lead and be involved in the selection process for both types of metrics:

- **Leadership**:

  - **Project or program managers** are typically at the forefront of leading the selection process for both outcome and progress metrics. They have a holistic view of a project's goals, scope, timeline, and resources, which positions them well to understand which metrics are most relevant and how they can be effectively implemented and tracked.

- **Team member involvement**:

  - **Executive sponsors**: Their involvement ensures that the selected metrics align with an organization's strategic objectives. Executives can provide the necessary support and resources, helping to integrate these metrics into broader business performance evaluations.

  - **Business analysts**: These professionals help translate business objectives into measurable metrics. They ensure that both outcome and progress metrics reflect the value and impact the project should deliver to the business.

  - **Technical leads and architects**: For technology-driven projects, their technical expertise is vital in determining the feasibility of capturing specific metrics. They can advise on the tools and systems needed to track these metrics effectively.

  - **QA managers**: Involving QA managers is crucial for defining metrics related to product or service quality. They contribute to selecting progress metrics that monitor ongoing quality controls and outcome metrics that reflect the final quality standards.

  - **Finance representatives**: They can provide insight into how the outcomes and the progress of the project align with financial goals, offering metrics that may relate to cost efficiency, **return on investment** (**ROI**), and budget adherence.

  - **Marketing and sales teams**: For projects that impact product offerings, involving these teams can help align metrics with customer satisfaction, market response, and sales performance. This ensures that the metrics chosen will help measure market success and customer engagement.

  - **Operations managers**: Particularly for projects impacting operations, these managers ensure that the metrics reflect operational efficiencies, process improvements, and service delivery standards.

  - **IT and data analysts**: Their role is essential in defining and setting up systems for data collection, analysis, and reporting selected metrics. They help ensure that data-driven metrics are accurate, timely, and relevant.

A collaborative approach to selecting metrics helps to cover all bases, from strategic alignment to practical measurement and day-to-day impact. This diversity in perspectives ensures that the metrics chosen are comprehensive and provide meaningful insights into both the progress of a project and its final outcomes.

Practices for designing metrics and dashboards

This section will give you practical advice on how to present data effectively so that all stakeholders can make quick, informed decisions.

To effectively track and visualize outcome and progress metrics in areas such as continuous testing, quality, security, and feedback, organizations need a thoughtful approach to designing these metrics and the dashboards that will display them. *Figure 12.14* illustrates the recommended process for selecting metrics.

Designing Metrics
1. Define measurement goals
2. Establish a baseline and targets
3. Choose data sources and collection methods
4. Implement data collection
5. Validate and refine metrics

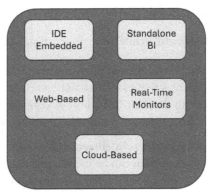

Figure 12.14 – Designing metrics and dashboards

Here's a breakdown of the process for designing these metrics and an exploration of the different architectures for the dashboards.

Designing an outcome and progress metrics

Here is the process for designing metrics:

1. **Define the measurement criteria**: Once metrics have been selected based on strategic goals, define specific measurement criteria and standards. This includes setting clear definitions for each metric measure, how data will be collected, and the frequency of measurement. For instance, if the metric is "the time to detect security incidents," determine whether this starts when the incident occurs or when it's first noticed by systems or personnel.

2. **Establish baselines and targets**: For each metric, establish a baseline that reflects current performance and set realistic targets for improvement. This provides a starting point to measure progress and a clear goal that drives actions. Baselines can be historical data, industry standards, or benchmarks.

3. **Choose data sources and collection methods**: Identify where and how data for each metric will be collected. For continuous testing, data might come from testing tools and CI/CD pipelines; for security metrics, data could be sourced from security monitoring and incident management systems.

4. **Implement data collection**: Ensure that the infrastructure and tools needed for automatic data collection are in place. This might involve configuring software tools, setting up APIs, and ensuring data accuracy and consistency.

5. **Validate and refine metrics**: Periodically review the metrics to ensure they provide valuable insights, and refine them based on feedback and changing objectives. This might involve adjusting data sources, collection frequency, or even the metric definitions themselves.

Architectures for dashboards displaying metrics

Alternative dashboard architectures, as illustrated in *Figure 12.14*, are as follows:

- **Integrated Development Environment (IDE)-embedded dashboards**: For metrics related to development activities such as continuous testing or code quality, dashboards can be integrated directly into the IDE. This allows developers to see metrics in real time within their working environment, enhancing immediacy and relevance.

- **Standalone Business Intelligence (BI) tools**: Tools such as Tableau, Power BI, or custom-built dashboard solutions can be used to pull data from various sources into a central dashboard. These tools offer powerful data visualization capabilities and can combine data from multiple areas, such as testing, security, and customer feedback, into a comprehensive overview.

- **Web-based dashboards**: Implementing web-based dashboards using frameworks such as Django (Python), Flask, or Angular, or even JavaScript libraries such as React, allows for highly customizable and accessible dashboards. These can be accessed via any web browser, providing flexibility and central access to stakeholders across an organization.

- **Real-time monitoring dashboards**: For metrics that require real-time monitoring, such as security incident detection or live feedback collection, dashboards built on platforms such as Grafana or Kibana can be used. These tools can integrate with databases and monitoring systems to provide live updates and alerts.

- **Cloud-based dashboards**: Cloud platforms such as Amazon CloudWatch, Google Cloud's operations suite, or Azure Monitor provide built-in tools to create dashboards that are scalable and accessible from anywhere. These are particularly useful for organizations with cloud-first strategies, offering integrated security and maintenance.

By following these processes and considering various architectural approaches for dashboards, organizations can ensure they have effective and efficient ways to monitor, analyze, and act on outcome and progress metrics across different domains. This enhances decision-making, improves strategic alignment, and drives operational efficiency.

Sustaining measures of progress and outcomes

Sustaining measures of outcomes and progress in an organization is an ongoing process that requires regular evaluation and adjustment, ensuring that metrics remain relevant and effective. Here are some recommended approaches for maintaining, evaluating, updating, and validating these metrics:

Evaluating and deprecating metrics

- **Regular review cycles**: Establish regular intervals (e.g., quarterly or biannually) to review the effectiveness and relevance of each metric. These reviews should assess whether the metrics are still aligned with current business objectives and strategies.

- **Performance analysis**: Evaluate metrics based on their ability to influence decision-making and business outcomes. If a metric consistently fails to provide actionable insights or drive improvements, it may need to be revised or deprecated.

- **Stakeholder feedback**: Gather feedback from users of the metrics, such as managers, team leaders, and analysts, to understand their practical utility and any challenges in using them. If metrics are found to be confusing, misaligned, or redundant, consider modifying or removing them.

- **Trigger events for deprecation**: Define specific criteria or trigger events that would warrant the deprecation of a metric. This could include significant changes in business processes, shifts in strategic focus, or the introduction of new technologies that make older metrics obsolete.

Introducing new metrics

- **Gap analysis**: Regularly perform gap analyses to identify areas where existing metrics do not fully cover key performance aspects. This helps to identify the need for new metrics.

- **Pilot testing**: Before fully implementing new metrics, conduct pilot tests to assess their relevance and effectiveness. This can be done in a controlled segment of an organization to minimize disruption and gather insightful data on a metric's performance.

- **Stakeholder buy-in**: Ensure that stakeholders understand the value of new metrics and how they contribute to an organization's goals. This includes training sessions, demonstrations, and discussions on how these metrics help in decision-making processes.

Validating metric implementations

- **Version management**: Treat metrics and their definitions as you would any software asset. Use version control systems to manage changes over time. This ensures that any modifications to metrics are well-documented and traceable.

- **Regression testing**: Each time a metric is updated or a new metric is introduced, perform regression testing to ensure that these changes do not adversely affect the accuracy and reliability of other metrics or the data collection process itself. This is critical to maintaining the integrity of data analytics.

- **Continuous validation**: Implement continuous validation mechanisms to monitor the performance of metrics. This could involve automated alerts that notify stakeholders when data patterns deviate significantly from historical trends, indicating potential issues with the metric's implementation.

- **Documentation and transparency**: Maintain comprehensive documentation for each metric, including its definition, purpose, calculation method, data source, and any changes made over time. This transparency helps build trust in the metrics and facilitates easier onboarding and training of new team members.

By following these approaches, organizations can ensure that their outcome and progress metrics remain relevant, reliable, and aligned with their strategic goals. This dynamic approach to metric management helps organizations adapt to changing environments and maintain their competitive edge by making informed, data-driven decisions.

Summary

This chapter explained the critical role of measuring progress and outcomes within organizations, with a focus on continuous testing, quality, security, and feedback. It began by emphasizing the importance of establishing effective measures to track how well organizations performed and adapted as they grew more complex. The chapter explained the fundamentals of different kinds of measurements, providing a basis for understanding what makes an effective metric and how metrics should align with an organization's goals.

As the chapter progressed, it offered detailed guidance on selecting the right metrics for both outcomes and progress. This section highlighted the importance of distinguishing between metrics that simply provide data and those that can truly drive transformative changes within processes.

We then offered practical advice on implementing these metrics through well-designed metrics systems and dashboards. This included how to present data in a way that is actionable and accessible for all stakeholders, enabling them to make informed decisions quickly.

The chapter concluded by addressing the sustainability of these measurements. It outlined strategies to maintain their relevance and usefulness over time, ensuring that they continue to support an organization's evolving needs. The next chapter will build on this by exploring emerging trends in continuous testing, quality, security, and feedback, offering insights into how these areas are likely to evolve and how organizations can adapt to these changes.

Part 4: Exploring Future Trends and Continuous Learning

Part 4 of this book delves into the evolving landscape of continuous practices in software development. This section begins by identifying emerging trends that are reshaping how continuous testing, quality, security, and feedback are integrated within software development frameworks. It provides insights into the latest advancements and how they can be leveraged to enhance operational maturity in organizations.

Following the discussion on emerging trends, the book shifts focus to the importance of continuous improvements and learning. It outlines effective strategies for fostering an environment of ongoing development and refinement in the areas of testing, quality, security, and feedback. This part of the book emphasizes the need for organizations to adapt and evolve continuously to keep pace with technological changes and to maintain a competitive edge in their respective industries.

This part includes the following chapters:

- *Chapter 13, Emerging Trends*
- *Chapter 14, Exploring Continuous Learning and Improvement*

13

Emerging Trends

This chapter describes the emerging trends that are reshaping the landscape of continuous testing, quality, security, and feedback within software development. As technology evolves and the pace of digital transformation accelerates, organizations encounter new challenges and opportunities that demand innovative responses. This chapter will explore how changes in adoption rates, the integration of advanced frameworks, and the rise of new methodologies such as observability, value stream management, platform engineering, and **artificial intelligence (AI)**/**machine learning (ML)** are influencing these areas. These trends are not just reshaping tools and processes but also redefining how teams collaborate and deliver value.

The chapter is structured to provide a comprehensive overview of these developments across five key areas – macro trends in DevOps, DevSecOps, and SRE; testability and observability; platform engineering; value stream management; and AI/ML. Each section will detail how these trends manifest in the real world, providing insights into their practical implications and the benefits they offer. For example, the shift toward integrating security practices within DevOps – forming DevSecOps – enhances security measures without compromising the speed of development cycles.

In addition to understanding these trends, the chapter aims to equip you with the knowledge to effectively prepare for and adapt to these changes. It's crucial for professionals in the field to not only grasp the theoretical aspects of these trends but also understand how to implement them, enhancing efficiency, quality, and security in their projects. This chapter will include discussions on strategic adjustments, skills enhancement, and adopting new tools and technologies that align with these emerging trends.

By the end of this chapter, you will have a solid understanding of the significant trends currently shaping the domains of continuous testing, quality, security, and feedback. You will learn how to leverage these trends to stay ahead in the rapidly evolving tech landscape, ensuring that your skills and methods are up to date and effective in addressing both current and future challenges in software development. This knowledge will be critical as you work to enhance your capabilities and adapt your strategies to meet the demands of modern software delivery practices.

In this chapter, we will cover the following topics:

- Macro trends in DevOps, DevSecOps, and SRE

- Testability and observability trends

- Platform engineering trends

- Value stream management trends

- AI/ML trends

Let's get started!

Macro trends in DevOps, DevSecOps, and SRE

The landscape of software development and operations is constantly evolving, influenced by technological advancements, changing market demands, and the need for faster, more secure delivery cycles. As we look into the future of continuous testing, quality, security, and feedback within the frameworks of DevOps, DevSecOps, and **Site Reliability Engineering (SRE)**, several macro trends stand out, as illustrated in *Figure 13.1*.

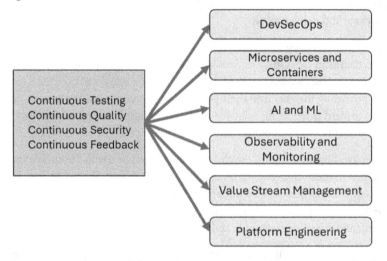

Figure 13.1 – Emerging trends for continuous testing, quality, security, and feedback

The following explains how each of these trends is relevant to the evolution of continuous testing, quality, security, and feedback:

- **A shift toward DevSecOps**: The integration of security practices into the DevOps pipeline – referred to as DevSecOps – is becoming a significant trend. As cyber threats increase in sophistication, organizations are recognizing the importance of embedding security early and throughout the software development life cycle, rather than treating it as an afterthought. This trend emphasizes automated security scanning tools, secure coding practices, and continuous compliance monitoring, all aimed at reducing vulnerabilities without slowing down development.

- **The increased adoption of microservices and containerization**: Microservices architecture and containerization facilitate the rapid, reliable delivery of large, complex applications. This trend impacts continuous testing, quality, and feedback by necessitating tools and practices that can handle more dynamic, distributed, and scalable environments. Tools such as Docker and Kubernetes have become standard, pushing developments in testing frameworks that can operate at scale and in real time to ensure quality and performance in microservices environments.

- **The growth of AI and ML**: AI and ML are increasingly being employed to enhance various aspects of DevOps, especially in continuous testing and feedback mechanisms. AI-driven predictive analytics can forecast potential bottlenecks and failures before they occur, while ML algorithms are used to optimize testing processes, automate error diagnoses, and refine feedback loops. These technologies allow teams to preemptively address issues, enhancing both the speed and quality of deliverables.

- **An emphasis on observability and monitoring**: As systems grow in complexity, so does the need for sophisticated monitoring and observability. This trend involves the transition from traditional monitoring to a more holistic approach known as observability, which focuses on understanding the internal states of systems by analyzing external outputs. It is particularly crucial in continuous feedback and operational resilience, enabling SRE teams to predict and mitigate disruptions before they affect service quality.

- **The rise of value stream management**: **Value Stream Management** (**VSM**) is gaining traction as organizations strive to optimize their end-to-end software delivery process to deliver maximum value. VSM tools and practices help map out the entire software delivery life cycle, identify inefficiencies, and measure the impact of improvements. This holistic view supports better decision-making in continuous testing, quality assurance, security practices, and feedback cycles, ensuring that efforts are aligned with business goals.

- **Platform engineering**: As a response to the complexity of managing microservices and cloud-native technologies, there is a growing trend toward platform engineering. This approach involves creating self-service platforms for developers that abstract away much of the complexity of infrastructure management. This enables developers to focus more on coding while relying on the platform to handle aspects of testing, security, and operations.

These trends highlight the dynamic nature of software development and operations, pointing to a future where adaptability, security, and efficiency are more interconnected than ever. For organizations and professionals in DevOps, DevSecOps, and SRE, staying abreast of these trends is crucial for maintaining competitive advantage and ensuring the delivery of high-quality, secure software products.

Testability and observability trends

The rising need for improved testability and observability is transforming the practices of continuous testing, quality, security, and feedback within the realms of DevOps, DevSecOps, and SRE. These two aspects are critical in creating more resilient, efficient, and secure applications and systems. Here's how their increasing significance is likely to impact these areas:

- **The impact on continuous testing and quality**:

  - **Enhanced test coverage and efficiency**: Improved testability means that systems are designed from the ground up and are easier to test. This can involve architectural decisions that favor modular designs or the incorporation of tools that facilitate automated testing. As a result, testing can be more thorough, less time-consuming, and more integral to the development process, leading to higher overall software quality.

  - **Faster feedback loops**: Observability provides deeper insights into how software behaves in production, which feeds back into the testing phase. By understanding a system's internal state better through metrics, logs, and traces, teams can identify not only what went wrong but also why. This enables faster debugging, more precise tests, and quicker iterations, all of which enhance the speed and efficacy of development cycles.

- **The impact on security in DevSecOps**:

  - **Proactive security measures**: Observability tools allow for real-time monitoring of security threats, enabling teams to detect and respond to issues as they arise, rather than afterward. This shift from reactive to proactive security is crucial in maintaining the integrity and safety of systems in a landscape of increasingly sophisticated cyber threats.

  - **Security as a shared responsibility**: As testability improves, security tests can become more integrated into the regular testing regime, blurring the lines between developers, testers, and security professionals. This integration helps foster a culture of "security as code," where security measures are baked into the software development life cycle from the beginning.

- **The impact on SRE**:

  - **Improved reliability and uptime**: Observability is a cornerstone of SRE, as it provides the data necessary to meet reliability targets and **service level objectives** (**SLOs**). Enhanced observability means better operational visibility, which helps in preemptively addressing potential downtimes and failures, thus improving system reliability and uptime.

  - **Data-driven operations**: With better observability tools, SREs can leverage extensive datasets to automate responses to common scenarios, predict system behavior under various conditions, and optimize resource usage. This not only reduces the burden of manual oversight but also allows SREs to focus on more strategic initiatives that add value.

- **Broader implications for DevOps, DevSecOps, and SRE:**

  - **A cultural shift toward data-driven decision-making**: As observability and testability improve, organizations can shift toward a culture that prioritizes data-driven decisions. This shift enhances not just technical processes but also strategic planning and resource allocation.

  - **Increased adoption of AI and ML**: The extensive data generated from improved testability and observability can be leveraged using AI and ML to further automate testing, threat detection, anomaly detection, and problem resolution processes. This can significantly speed up development cycles, enhance security postures, and improve system resilience.

In conclusion, a focus on improving testability and observability is pivotal in addressing the complexities of modern software development and operations. As these trends continue to evolve, they are likely to bring about significant enhancements in testing, quality assurance, security, and operational reliability, fundamentally changing how teams approach DevOps, DevSecOps, and SRE practices.

Platform engineering trends

The rising need for platform engineering solutions is fundamentally reshaping the landscape of DevOps, DevSecOps, and SRE by providing robust, scalable, and efficient infrastructure to support continuous testing, quality assurance, security, and feedback mechanisms. Platform engineering focuses on creating shared, self-service platforms for developers and operators that abstract away the complexities of underlying infrastructure, allowing teams to focus more on application development and less on the operational details. Here's how this shift is likely to impact key areas:

- **Continuous testing and quality:**

  - **Streamlined testing processes**: Platform engineering solutions typically integrate and standardize testing tools and environments, which can dramatically streamline the testing process. By providing pre-configured environments and automated pipelines, these platforms enable consistent testing procedures, reduce setup time, and eliminate the "works on my machine" problem, thereby enhancing the quality of software.

  - **Enhanced test automation**: With platform engineering, organizations can implement more sophisticated test automation strategies that are scalable and repeatable across various development projects. This not only speeds up the testing cycles but also improves the coverage and reliability of tests, leading to higher software quality.

- **Security in DevSecOps:**

  - **Integrated security practices**: Platform engineering promotes the integration of security tools directly into the development and deployment pipelines. By embedding security as a core component of the platform, it ensures that security checks and controls are automatically applied throughout the software development life cycle, enhancing the security posture without adding overhead for developers.

- **Consistency and compliance**: Since platform engineering standardizes the development and deployment workflows, it also standardizes security practices. This consistency is crucial for maintaining security compliance across all applications and services, particularly in regulated industries where consistency in security implementations can simplify compliance audits.

- **SRE teams**:

  - **Improved operational efficiency**: Platform engineering empowers SRE teams by automating many operational tasks related to deployment, monitoring, and scaling. This automation frees up SREs to focus on higher-value activities, such as improving system architecture and optimizing performance, rather than getting bogged down in routine operational issues.

  - **Proactive monitoring and reliability**: Platforms engineered with observability in mind provide SRE teams with the tools to monitor applications and infrastructure proactively. This capability allows for quicker detection and resolution of issues, often before they impact users, thus improving the overall reliability and uptime of services.

- **Feedback mechanisms**:

  - **Faster feedback loops**: By standardizing and automating the collection and analysis of feedback from both users and internal monitoring tools, platform engineering solutions can significantly shorten the feedback loop. This rapid feedback is crucial for iterative development and continuous improvement practices in DevOps.

  - **Data-driven decision making**: The centralized nature of platform-engineered solutions allows for more comprehensive data collection across the entire software life cycle. This wealth of data enables more informed decision-making, helping teams prioritize changes and improvements based on real user feedback and system performance data.

- **Broader implications**:

  - **A cultural shift toward automation and self-service**: Platform engineering encourages a shift toward automation and self-service capabilities. This shift not only reduces the cognitive load on developers and operators but also fosters a culture of innovation and experimentation, by making it easier to test new ideas in a controlled, reproducible manner.

  - **Enhanced collaboration**: By reducing the friction associated with development and operational processes, platform engineering can lead to better collaboration between development, operations, and security teams. This enhanced collaboration is key to the DevOps philosophy and crucial for the rapid delivery of high-quality software.

In conclusion, the rising need for platform engineering solutions is likely to provide significant benefits in terms of efficiency, quality, security, and operational reliability across DevOps, DevSecOps, and SRE. As these platforms become more sophisticated, they will continue to drive the evolution of software development and operations practices, making them more agile, secure, and user-centric.

VSM trends

The rising need for VSM solutions is making a significant impact on continuous testing, quality, security, and feedback within the frameworks of DevOps, DevSecOps, and SRE. VSM focuses on visualizing and optimizing the flow of value from idea to delivery, offering a systematic approach to improve efficiency and effectiveness throughout the software development life cycle. Here's how VSM is poised to enhance these areas:

- **Continuous testing**:

  - **Enhanced visibility into testing phases**: VSM provides a clear visualization of each phase of the development process, including all testing stages. By mapping out when and how testing occurs within the value stream, teams can identify bottlenecks or inefficiencies in the testing process. This visibility allows for targeted improvements, such as the integration of automated testing tools at critical points to speed up the workflow without sacrificing quality.

  - **Optimized resource allocation**: With a better understanding of the testing process through VSM, organizations can more effectively allocate resources – both human and technical – to ensure that testing does not become a bottleneck. This can lead to a more balanced distribution of tasks and reduce wait times for test environments or necessary approvals.

- **Quality**:

  - **Improved feedback for quality assurance**: VSM tools often integrate feedback mechanisms that can track the quality metrics throughout the software life cycle. This integration allows for immediate feedback on quality issues, which can be addressed more rapidly and with greater precision. Continuous improvement in quality is facilitated by real-time data, which helps in fine-tuning the processes continually.

  - **Consistency across the value stream**: VSM encourages standardization of processes, which in turn helps maintain a consistent level of quality across the entire stream. By understanding how changes in one part of the stream affect downstream outcomes, teams can implement quality checks at strategic points to ensure that the final product meets the desired standards.

- **Security in DevSecOps**:

  - **Proactive security integration**: VSM provides a framework for integrating security practices throughout the development life cycle, not just at the end. By including security as a part of the value stream, it becomes an integral component of the product development, ensuring that security considerations are made early and often. This early integration helps in identifying vulnerabilities sooner and reduces the cost and complexity of fixes.

  - **Enhanced traceability**: VSM solutions improve traceability across the value stream. This capability is crucial for security, as it allows teams to quickly identify the source of security issues and understand their impact across the entire stream. Effective traceability supports better risk management and compliance tracking, essential aspects of secure software delivery.

- **Feedback:**

  - **Accelerated feedback loops**: VSM tools streamline the capture and utilization of feedback from various stages of the development process. By linking feedback directly to specific stages in the value stream, teams can more quickly and accurately make adjustments that enhance a product's value to users. This rapid integration of feedback is crucial for agile development practices and helps ensure that the final product aligns with user needs and expectations.

  - **Data-driven improvements**: The analytics capabilities of VSM allow organizations to gather data from across the value stream and use it to make informed decisions about where improvements are needed. This data-driven approach to managing feedback helps prioritize efforts based on impact, ensuring that resources are focused on areas that will provide the greatest benefit.

In conclusion, the implementation of VSM solutions is revolutionizing the way organizations approach continuous testing, quality, security, and feedback. By providing a holistic view of the value creation process, VSM enables more strategic decision-making, enhances operational efficiency, and improves the overall quality and security of software products. For DevOps, DevSecOps, and SRE teams, embracing VSM can lead to significant gains in productivity, security, and customer satisfaction.

AI/ML trends

The integration of AI and ML into DevOps, DevSecOps, and SRE practices is rapidly transforming how teams approach continuous testing, quality, security, and feedback. The rising need for AI/ML solutions is driven by the increasing complexity of systems, the volume of data generated, and the demand for faster, more efficient delivery cycles. Here's how AI/ML is likely to impact these crucial areas:

- **Continuous testing:**

  - **Automated test creation and optimization**: AI/ML can analyze application data and user interactions to automatically generate and optimize test cases, reducing the manual effort required in test creation. This leads to broader test coverage and more efficient testing processes, ensuring that all relevant application scenarios are accounted for.

  - **Predictive analytics for flaw detection**: ML algorithms can predict potential flaws based on historical data, enabling teams to focus testing efforts where they are most needed before defects become apparent. This proactive approach not only saves time but also reduces the cost associated with late-stage bug fixes.

- **Quality:**

  - **Enhanced error detection**: AI algorithms excel at identifying patterns and anomalies. In quality assurance, these capabilities can be harnessed to detect deviations from expected outcomes or identify unusual system behavior that might indicate underlying quality issues. This early detection helps to maintain high quality throughout the development cycle.

- **Dynamic adaptation to changes**: AI-driven systems can dynamically adapt testing and quality assurance processes based on ongoing changes in the code base and operational environment. This responsiveness ensures that quality checks remain relevant and effective even as a project evolves.

- **Security in DevSecOps**:

  - **Intelligent threat detection and response**: AI/ML models can analyze vast amounts of security data in real time to identify potential threats that might be missed by traditional security tools. These models learn from new data, continuously improving their detection capabilities. Additionally, AI can automate responses to common threats, enabling faster mitigation and reducing the workload on security teams.

  - **Security as code**: AI enhances the concept of security as code, where security policies and checks are integrated into the development pipeline. AI can help to fine-tune these policies by learning system behaviors and threat patterns, ensuring robust security protocols that evolve with the threats.

- **Feedback**:

  - **Real-time feedback analysis**: AI/ML can process and analyze feedback from users and systems in real time, providing actionable insights more quickly. This capability allows development teams to make informed decisions rapidly and to iterate on products in a way that closely aligns with user needs and expectations.

  - **Sentiment analysis and user experience**: ML models can perform sentiment analysis on user feedback to gauge the emotional tone, providing deeper insights into user satisfaction and potential pain points. This analysis helps to prioritize development efforts based on the impact on user experience.

- **The overall impact on DevOps, DevSecOps, and SRE**:

  - **Automated decision-making**: AI enhances decision-making processes by providing data-driven insights across all stages of the software development life cycle. This capability supports more accurate and quicker decisions regarding testing, deployment, operations, and resource allocation.

  - **Enhanced efficiency and innovation**: By automating routine tasks and analyses, AI/ML frees up human resources to focus on more complex and innovative work. This shift not only boosts efficiency but also fosters creativity within teams, leading to more innovative solutions and advancements.

  - **Scalability and handling complexity**: AI/ML solutions are inherently scalable and capable of handling the increased complexity of modern software environments. This scalability is crucial for maintaining performance and quality as systems grow and evolve.

In conclusion, the rise of AI/ML in DevOps, DevSecOps, and SRE is set to profoundly influence continuous testing, quality assurance, security practices, and feedback mechanisms. As these technologies become more integrated into everyday practices, they promise to enhance the speed, efficiency, and effectiveness of software development and operations, ultimately leading to more robust, secure, and user-aligned products.

Summary

This chapter explored the emerging trends affecting continuous testing, quality, security, and feedback within the realms of DevOps, DevSecOps, and SRE. The chapter began by detailing macro trends, such as the shift toward DevSecOps, the adoption of microservices and containerization, the growth of AI and ML, the emphasis on observability and monitoring, and the rise of VSM. Each trend was analyzed for its impact on software development processes, highlighting how these advances facilitate faster, more secure delivery cycles, enabling organizations to better meet the demands of modern digital landscapes.

The discussion then delved into specific areas affected by these trends, including testability and observability, VSM, platform engineering, and AI/ML integration. The chapter explained how improved testability and observability enhance the efficiency and effectiveness of continuous testing and feedback mechanisms, allowing for quicker iterations and more reliable software outputs. It also covered how platform engineering is reshaping the way teams handle the complexities of modern software development by creating robust, scalable infrastructures that automate and streamline operations.

Furthermore, the chapter addressed the integration of AI and ML technologies, which are transforming testing, quality assurance, and security protocols. AI/ML enables predictive analytics for flaw detection, automated test case generation, and enhanced real-time monitoring and feedback analysis. These capabilities are leading to smarter, more proactive development environments where continuous improvement is ingrained in the workflow.

In conclusion, the chapter equipped you with the knowledge to understand and implement these emerging trends, thereby preparing you to adapt to the continuously evolving tech landscape. The next chapter will build on this, focusing on continuous learning for continuous testing, quality, security, and feedback for DevOps, DevSecOps, and SRE, aiming to further enhance the skills necessary to thrive in these dynamic fields.

14

Exploring Continuous Learning and Improvement

This chapter, as illustrated in *Figure 14.1*, focuses on how teams that are involved in DevOps, DevSecOps, and SRE can continually improve and learn, which is essential in fast-changing tech environments. The chapter explores effective strategies for continuous learning and improvement in areas crucial for software development and operations. The goal is to help teams not just keep up to date with technological changes but also use them as opportunities for advancement.

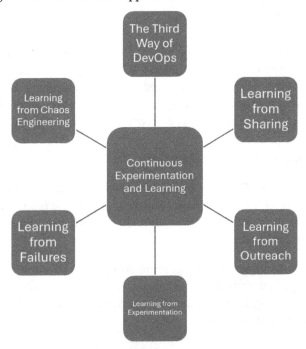

Figure 14.1 – Continuous experimentation and learning

The chapter is structured around the *Third Way of DevOps*, a principle that promotes constant learning through experimentation. It discusses various ways teams can learn – through sharing, outreach, experimenting, handling failures, and applying chaos engineering. Each section covers different aspects of learning, providing practical advice on how to integrate these practices into daily work routines.

By covering these topics, this chapter aims to equip you with the skills needed for experimentation and the application of chaos engineering in testing.

In this chapter, we'll cover the following main topics:

- The Third Way of DevOps
- Learning from sharing
- Learning from outreach
- Learning from experimentation
- Learning from failure
- Learning from chaos engineering

Let's get started!

The Third Way of DevOps

The Third Way of DevOps focuses on the principles of continual experimentation and learning to enhance the workflow within software development and operations. This approach not only encourages innovation but also emphasizes the importance of learning from successes and failures. It is particularly crucial for areas such as continuous testing, quality, security, and feedback, where rapid advancements and changes in technology demand adaptability and continuous skill enhancement.

Continuous improvement in DevOps

Continuous improvement in DevOps involves regularly evaluating and enhancing processes, tools, and methodologies used in software development and deployment. This practice ensures that teams do not maintain the status quo and, instead, actively seek ways to increase efficiency, effectiveness, and security. The key aspects include the following:

- **Iterative process enhancements**: DevOps teams are encouraged to constantly look for small, incremental changes in their workflows that could lead to significant improvements over time. This could be automating a routine manual process, refining an existing script, or adopting new tools that speed up deployment.

- **Feedback loops**: Creating quick and effective feedback loops is essential in DevOps. These loops help teams understand the outcomes of their changes in real time and adjust accordingly, without waiting for the end of a long project cycle. Feedback can come from automated systems, peer reviews, or real-time monitoring tools, and it should directly inform future improvements.

Learning in DevOps

Learning in DevOps is not just about acquiring new knowledge; it's also about applying it practically and sharing it with the entire team. This facet of the Third Way is critical for fostering a culture of collective intelligence where everyone's growth benefits the whole team.

- **From failures**: DevOps culture does not shun failures and, instead, sees them as opportunities to learn. When a deployment fails or a new bug is discovered, teams conduct blameless post-mortems to understand what went wrong and how similar issues can be avoided in the future. Documenting these lessons and sharing them ensures that the same mistakes are not repeated.

- **From successes**: Similarly, analyzing what worked well and why can provide valuable insights. Success stories should be dissected with the same rigor as failures to replicate these successful actions in future projects.

Continuous testing, quality, and security

The application of the Third Way in continuous testing, quality, and security involves several practices:

- **Continuous testing**: Automated testing is a cornerstone of DevOps, providing immediate feedback on the impact of changes. By continuously testing software from early in the development process, teams can detect and address issues sooner, which reduces costs and speeds up delivery times.

- **Quality assurance**: Quality is not an afterthought but is integrated throughout the development process. Teams use insights from past projects to enhance quality checkpoints and standards continually.

- **Security integration**: Integrating security into the DevOps process, known as DevSecOps, from the start ensures that security considerations are built-in and not bolted on. Learning from past security incidents helps to fortify future projects against similar vulnerabilities.

The Third Way of DevOps advocates for a culture where continuous improvement and learning are part of the daily routine. By embracing this approach, DevOps teams enhance their capabilities in continuous testing, quality, security, and feedback. This not only leads to more robust and resilient software solutions but also ensures that teams remain at the cutting edge of technology innovation, prepared to tackle new challenges as they arise. By integrating these practices, organizations can sustain long-term growth and continuous improvement in their software development endeavors.

Learning from sharing

Learning from sharing is a fundamental concept within DevOps that emphasizes the benefits of knowledge exchange within and between teams. In the context of continuous improvement and learning for continuous testing, quality, security, and feedback, sharing becomes a crucial driver for enhancing team capabilities and accelerating development cycles. This section explores how structured sharing practices can foster a collaborative environment where innovation thrives.

Building a culture of open communication

For DevOps teams, creating a safety culture where members feel comfortable sharing insights, successes, and failures is essential. This culture promotes transparency and trust, which are foundational for effective collaboration.

- **Regular knowledge-sharing sessions**: Implementing regular meetings such as daily stand-ups, weekly reviews, or monthly retrospectives where team members discuss what they are working on, the challenges faced, and the lessons learned can significantly enhance collective understanding and problem-solving capabilities.

- **Documentation**: Maintaining a knowledge base with detailed documentation of processes, incidents, and solutions not only serves as a reference but also helps new team members get up to speed quickly. Wikis, shared drives, and internal blogs can be effective tools for this purpose.

Sharing best practices and tools

DevOps teams often work in rapidly evolving environments where new tools and practices frequently emerge. Sharing best practices and tools within a team or organization can lead to more efficient and effective workflows:

- **Tool demonstrations**: Regular sessions to demonstrate new tools or features that team members have found useful can facilitate wider adoption and standardization across a team.

- **Best practice workshops**: Workshops that focus on the best practices for continuous testing, quality, and security can help standardize approaches across a team, reducing variability and increasing the reliability of outcomes.

Cross-team collaboration and external engagement

Extending a culture of sharing beyond immediate teams to other departments, and even outside the organization, can introduce fresh perspectives and solutions that internal teams might overlook.

- **Inter-departmental meetings**: Encouraging teams to hold cross-departmental meetings can help break down silos, allowing for the sharing of insights that could benefit multiple areas, from development to operations to security.

- **Community Involvement**: Participating in external forums, conferences, and workshops allows team members to not only gain insights from industry leaders but also share their own experiences and solutions, establishing a presence in the broader community.

Leveraging feedback for continuous improvement

Feedback mechanisms are integral to software processes, and sharing feedback across teams is vital for continuous improvement. Structured feedback can inform better practices and highlight areas that need attention:

- **Feedback loops**: Integrating feedback loops into the sharing process ensures that all team members are aware of the outcomes of their actions and can adjust their practices accordingly. These loops should be timely and constructive, focusing on both what went well and what could be improved.

- **Actionable insights**: Feedback should be actionable. Teams need to discuss the practical steps that can be taken to implement the insights gained through sharing. This can involve adjustments to testing protocols, updates to security measures, or enhancements to the quality assurance processes.

In DevOps, learning from sharing is not just about distributing information but about creating an ecosystem where continuous learning and improvement are part of the everyday process. By fostering an open, communicative, and collaborative culture, teams can significantly enhance their effectiveness and efficiency in continuous testing, quality, security, and feedback. This not only boosts project outcomes but also contributes to a more engaged and innovative team environment. As teams become more adept at sharing and applying collective knowledge, they set a foundation for sustained growth and continuous advancement in their DevOps practices.

Learning from outreach

Outreach in the context of DevOps involves engaging with external communities and industry groups to bring in new ideas and practices that can enhance internal processes. This section examines how learning from outreach can significantly contribute to continuous improvement in continuous testing, quality, security, and feedback within DevOps, DevSecOps, and SRE frameworks.

The role of external engagement in continuous improvement

External engagement allows teams to step beyond the confines of their immediate organizational environment to learn from the broader industry. This can include participating in industry conferences, contributing to open source projects, or engaging in professional networking groups:

- **Participation in conferences and workshops**: Attending industry conferences and workshops provides teams with exposure to cutting-edge practices and technologies. These events are valuable for learning about new tools, methodologies, and strategies that have been effective in similar environments. Teams can bring back these insights and evaluate their applicability to their own processes.

- **Engaging with online communities**: Online forums and communities such as GitHub, Stack Overflow, or specific DevOps forums are rich resources for problem-solving and innovation. By actively participating in these communities, team members can ask questions, get feedback on their approaches, and learn from the experiences of others facing similar challenges.

The benefits of industry collaboration

Collaborating with industry peers offers several advantages that directly translate into improved performance in testing, quality, security, and feedback mechanisms:

- **Cross-industry learning**: By interacting with professionals from different industries, DevOps teams can learn how various challenges are addressed in different contexts. This cross-industry learning can inspire innovative solutions that teams might not have considered previously.

- **Benchmarking against industry standards**: Outreach allows teams to benchmark their processes and performance against industry standards and best practices. This benchmarking can highlight areas for improvement and provide a clear target for future development efforts.

Implementing outreach learnings in DevOps practices

To effectively integrate learnings from outreach activities, teams need to follow a structured approach to assimilation and application.

- **Structured debriefing sessions**: After participating in external events or engagements, team members should hold debriefing sessions to share key learnings with their peers. These sessions should focus on potential applications of the new information and identifying actionable steps for implementation.

- **Pilot projects**: Implementing new ideas through small-scale pilot projects can help teams assess their effectiveness without risking broader system stability. These projects serve as practical tests for new concepts and allow for iterative refinement before wider rollout.

Learning from outreach is a vital component of continuous improvement in DevOps environments. By actively engaging with external sources of knowledge and bringing those insights back into an organization, teams can enhance their capabilities in continuous testing, quality, security, and feedback. This not only keeps the team updated with the latest industry developments but also fosters a culture of learning and openness to change. As a result, DevOps teams are better equipped to adapt to new challenges and continuously refine their practices to achieve higher efficiency and effectiveness in their projects.

Learning from experimentation

Experimentation is a core element of the DevOps philosophy, promoting an environment where trying new approaches is encouraged to discover more efficient and effective methods. This section focuses on how experimentation can drive continuous improvement in continuous testing, quality, security, and feedback processes within DevOps, DevSecOps, and SRE.

The importance of experimentation in DevOps

Experimentation allows teams to test hypotheses in real-world scenarios, providing invaluable insights that can lead to significant enhancements in software development and operational processes. In DevOps, experimentation is not just about technological innovation but also about finding new ways to improve team dynamics, workflows, and overall efficiency:

- **Iterative testing**: In continuous testing, experimentation can involve trying out new automated testing tools or techniques to see whether they improve the speed and accuracy of tests. For instance, teams might experiment with different types of automated regression tests to determine which one provides the best balance between speed and thoroughness.

- **Quality improvement trials**: Quality assurance can benefit from experimenting with new quality metrics or new methods of gathering and analyzing quality data. By testing these different approaches on small portions of a project, teams can gather data on their effectiveness without impacting the entire project.

Conducting safe experiments in DevOps

To ensure that experimentation leads to learning rather than disruption, it is crucial to conduct experiments in a controlled, measurable way:

- **The use of feature flags**: Implementing feature flags allows teams to test new features or changes in a live production environment with minimal risk. Feature flags can be used to roll out changes to only a small subset of users, enabling teams to gather feedback and performance data without impacting all users.

- **Canary releases**: Similar to feature flags, canary releases involve rolling out new features or changes to a small group of users before a full rollout. This method is particularly useful for testing in live environments and can provide valuable feedback on how changes perform under real-world conditions.

Learning from experimentation outcomes

The key to successful experimentation in DevOps is not just conducting experiments but also learning from them, regardless of whether they succeed or fail:

- **Analyzing results**: Every experiment should be followed by a detailed analysis of its results. This includes reviewing performance metrics, user feedback, and any other relevant data to determine an experiment's success or failure.

- **Sharing learnings**: The insights gained from each experiment should be documented and shared with the entire team. This practice helps spread knowledge and ensures that all team members can learn from the experiment, even if they were not directly involved in it.

- **Incorporating feedback into processes**: The final step in the experimentation cycle is to use the feedback and learnings to refine existing processes or develop new ones. This might involve adjusting testing protocols, updating quality standards, or modifying security measures.

Learning from experimentation is a powerful tool for continuous improvement in DevOps. By fostering an environment where experimentation is encouraged and safely managed, teams can continuously evolve their practices and processes. This ongoing refinement leads to higher efficiency, better quality, enhanced security, and more effective feedback mechanisms, ensuring that an organization remains adaptive and innovative in the face of changing technology and market conditions.

Learning from failure

In DevOps, learning from failures is an essential practice that drives continuous improvement across all stages of software development and operations. This concept is rooted in the belief that failures provide valuable lessons that, when properly analyzed and understood, can lead to significant enhancements in processes, including continuous testing, quality, security, and feedback mechanisms.

Embracing a no-blame culture

A crucial element in learning from failures is the establishment of a no-blame culture. This environment encourages team members to openly discuss mistakes and failures without fear of retribution or criticism:

- **Blameless postmortems**: After a failure, conducting a blameless postmortem is vital. The objective is to understand what happened, why it happened, and how it can be prevented in the future without pointing fingers at individuals. This approach helps identify system-level improvements rather than focusing on individual errors.

- **Transparency**: Teams are encouraged to document and share details about failures and their root causes with all members. This transparency not only promotes learning but also helps to build trust and collaboration within a team.

Practical steps to analyze failures

Analyzing failures effectively involves a structured approach to uncovering the root causes and understanding the broader implications for continuous improvement:

- **Root Cause Analysis (RCA)**: Techniques such as the "five whys" (https://www.mindtools.com/a3mi00v/5-whys) or the fishbone diagram (https://www.techtarget.com/whatis/definition/fishbone-diagram) can be employed to drill down to the underlying reasons for a failure. The goal is to go beyond superficial answers and understand the systemic issues that need addressing.

- **Actionable lessons learned**: Each analysis should result in actionable takeaways that can be implemented to improve processes. Whether adjusting testing protocols, refining deployment practices, or enhancing monitoring tools, the key is to convert insights into tangible improvements.

Integrating failures into continuous improvement cycles

Learning from failures should be an integral part of the continuous improvement cycles within DevOps practices:

- **Iteration on feedback**: Insights from failures should feed directly back into the DevOps life cycle. This could mean changes to the code, updates to security measures, or enhancements in the monitoring strategies used during continuous testing and operations.

- **Regular review sessions**: Regularly scheduled review sessions should be held to assess recent failures and the status of implemented changes based on previous failures. This ensures that the lessons are not only learned but also effectively applied.

The benefits of learning from failures

Learning from failures has multiple benefits that directly impact the effectiveness and efficiency of DevOps practices:

- **Improved resilience and reliability**: By continuously learning from past failures, teams can build more resilient and reliable systems. Each failure provides a unique opportunity to fortify a system against future issues.

- **Enhanced innovation**: A culture that does not fear failure is one that encourages innovation. Teams feel more comfortable experimenting with new ideas when they know that failures are seen as learning opportunities rather than setbacks.

Learning from failures is a cornerstone of continuous improvement in DevOps, directly impacting continuous testing, quality, security, and feedback processes. By fostering a culture that sees failure as an opportunity to learn and improve, organizations can enhance their operational processes and adapt more quickly to new challenges. This approach not only minimizes the recurrence of similar failures but also drives innovation and efficiency in a continuously evolving technological landscape.

Learning from chaos engineering

Chaos engineering is a disciplined approach to identifying failures before they become outages. By intentionally injecting faults into systems, DevOps teams can observe how their systems behave under stress and learn how to build more robust systems. This proactive approach is crucial for continuous testing, enhancing quality, bolstering security, and refining feedback mechanisms in software development and operations.

Implementing chaos engineering

The practice of chaos engineering involves several planned steps to ensure that it adds value without causing unnecessary disruption to production services:

1. **Define clear objectives**: Before initiating chaos experiments, it's important to have clear goals. Whether it's testing how a system recovers from a database failure or understanding the impact of a sudden increase in traffic, defining these objectives helps tailor the experiments effectively.

2. **Start in a controlled environment**: Initially conducting chaos experiments in a staging environment helps teams learn and iterate on their processes without affecting live users. This controlled setting allows you to adjust variables and understand the outcomes without the pressure of immediate business impact.

3. **Gradual escalation**: Once confidence is built into controlled environments, gradually escalating chaos experiments to production ensures that systems are resilient under real-world conditions. This step-by-step approach helps mitigate risks while still gaining valuable insights.

Learning and improvement from chaos engineering

Chaos engineering provides critical learnings that can directly influence various aspects of DevOps processes:

- **Enhanced system resilience**: By breaking things on purpose, teams can identify and fix issues before they impact users. This proactive fault-finding helps improve a system's resilience and reduces downtime, which is crucial for maintaining high-quality user experiences.

- **Improved recovery processes**: Chaos experiments often reveal how well (or poorly) systems recover from failures. By learning from these outcomes, teams can streamline and strengthen their recovery procedures, such as improving automated rollbacks or enhancing alerting mechanisms.

- **Security posture assessment**: Introducing security-related chaos experiments, such as testing responses to simulated attacks, can help uncover vulnerabilities in the security posture. These insights drive improvements in security protocols and defense strategies.

Integrating chaos engineering into continuous feedback loops

To fully leverage the benefits of chaos engineering, it's essential to integrate the findings into the continuous feedback loops of DevOps practices:

- **Documentation and analysis**: After each chaos experiment, documenting the findings and conducting thorough analyses to extract actionable insights is crucial. This documentation should be accessible to all team members to ensure that everyone learns from the experiments.

- **Feedback mechanisms**: Integrating chaos engineering findings into existing feedback mechanisms helps keep all stakeholders informed about potential and existing weaknesses in a system. This ongoing communication is vital for continuous improvement.

- **Routine practice**: Making chaos engineering a routine part of the development life cycle normalizes the practice of learning from failures. Regularly scheduled chaos experiments ensure continuous learning and system hardening, which are critical for adapting to new challenges and technologies.

Chaos engineering is a powerful tool in the DevOps, DevSecOps, and SRE toolkits, offering a unique way to proactively test and improve systems. By intentionally introducing failures, teams can ensure their software and infrastructure can withstand unexpected disruptions. This learning method not only enhances the reliability and security of systems but also fosters a culture of resilience and preparedness within an organization. As teams become more comfortable with chaos experiments, they can continue to push the boundaries of what their systems can handle, leading to continuous improvement in all facets of software delivery and operations.

Summary

The final chapter of this book explored the critical practice of continuous learning and improvement within the contexts of continuous testing, quality, security, and feedback for DevOps, DevSecOps, and SRE. The chapter highlighted the importance of embracing a culture of continuous learning, emphasizing how it underpins success in rapidly evolving technological environments. The narrative was structured around several key methodologies that foster this continuous improvement, including the Third Way of DevOps, which advocates for a culture of feedback, learning from experimentation, and the pivotal role of recovering from failures.

We delved into specific strategies that facilitate learning and improvement, starting with the Third Way of DevOps. This approach was discussed in the context of fostering an environment that encourages experimentation and learning from successes and failures alike. It laid out practical steps to implement this philosophy within teams, ensuring that learning is embedded in the daily workflow and contributes directly to operational success.

Further sections of the chapter focused on learning through various lenses – sharing, outreach, experimentation, failures, and chaos engineering. Each section provided concrete examples of how these elements can be integrated into DevOps practices to enhance outcomes. For instance, learning from sharing emphasized the importance of knowledge exchange both within teams and with the external community, while learning from failures and chaos engineering stressed the value of systematic and intentional testing and breaking systems to improve their robustness and security.

Additionally, the importance of outreach and experimentation was underscored, showing how engagement with wider communities and proactive experimentation can lead to significant improvements in processes and systems. These sections collectively painted a comprehensive picture of how continuous learning and adaptation are not just beneficial but also essential, to maintain relevance and efficiency in the face of changing technology and market demands.

Thank you for taking the time to explore continuous testing, quality, security, and feedback for DevOps, DevSecOps, and SRE throughout this book. We hope the discussions and methodologies presented across the chapters have equipped you with valuable insights and practical knowledge to enhance your practices. Your engagement and commitment to learning are what drive improvement and innovation in this field. We appreciate your dedication to advancing your understanding and skills, and we wish you success in your journey of continuous learning and improvement in the future.

Glossary and References

Glossary of terms

Here's an expanded list of relevant glossary terms focused on continuous testing, quality, security, and feedback, organized alphabetically:

A

Acceptance testing: Testing that verifies a system meets agreed-upon requirements at the end of the development process and prior to release.

Automated build: The process of automating the compilation and construction of software code into executable software.

B

Behavior-driven development (BDD): A collaborative approach to software development and testing where business interests and technical solutions are aligned using shared tools and processes.

Blameless postmortems: A method used in DevOps to analyze and learn from failures without personal blame, focusing on process and system improvements.

C

Change management: The process, tools, and techniques to manage the people side of change to achieve the required business outcome.

Configuration management (CM): The technical and administrative activities concerned with the creation, maintenance, controlled change, and quality control of the scope of work.

Continuous delivery (CD): A software engineering approach in which teams produce software in short cycles, ensuring that the software can be reliably released at any time.

Continuous deployment: The practice of releasing software to production as soon as it has passed automated tests.

Continuous improvement: Ongoing effort to improve products, services, or processes. These efforts can seek "incremental" improvement over time or "breakthrough" improvement all at once.

Continuous integration (CI): A development practice where developers integrate code into a shared repository frequently, and each check-in is then verified by an automated build, allowing teams to detect problems early.

Continuous security: Integrating security practices and tools into the development and deployment pipelines to identify and mitigate risks continuously.

Continuous testing: The process of executing automated tests as part of the software delivery pipeline to obtain immediate feedback on the business risks associated with a software release candidate.

D

DevOps: A set of practices that combines software **development (Dev)** and IT **operations (Ops)** aimed at shortening the systems development life cycle and providing CD with high software quality.

DevSecOps: An augmentation of DevOps that incorporates security practices and principles into the DevOps process.

F

Feedback loop: A system in which outputs are circled back and used as inputs, central to continuous development, delivery, and improvement processes.

I

Incident management: The process of managing the life cycle of all incidents to ensure that normal service operation is restored as quickly as possible and the business impact is minimized.

M

Monitoring: The activity of observing and checking the behavior and outputs of a system and its components over time.

Mutation testing: A method of software testing that involves modifying programs' source code or byte code in small ways to test sections of the code base that are not found by existing tests.

P

Performance testing: Testing performed to determine how a system performs in terms of responsiveness and stability under a particular workload.

Postmortem analysis: The process after an incident or project completion to investigate and draw lessons for future improvement.

Q

Quality assurance (QA): Ensuring that a software product meets required quality standards before it is released into production.

Quality control (QC): The operational techniques and activities, part of quality management, focused on fulfilling quality requirements.

R

Regression testing: A type of software testing that verifies that software previously developed and tested still performs correctly after it was changed or interfaced with other software.

Risk management: The forecasting and evaluation of financial risks together with the identification of procedures to avoid or minimize their impact.

S

Security information and event management (SIEM): An approach to security management that seeks to provide a holistic view of an organization's information security.

Software testing: The process of executing a program or application with the intent of finding software bugs.

T

Test automation: The use of special software to control the execution of tests and the comparison of actual outcomes with predicted outcomes.

Threat modeling: A procedure for optimizing network security by identifying objectives and vulnerabilities, and then defining countermeasures to prevent, or mitigate the effects of, threats to the system.

U

Usability testing: Testing to evaluate a product by testing it on users to directly observe how they interact with it.

V

Vulnerability assessment: The process of identifying, quantifying, and prioritizing the vulnerabilities in a system.

Book references

- *Accelerate: The Science of Lean Software and DevOps: Building and Scaling High Performing Technology Organizations* by Nicole Forsgren, Jez Humble, and Gene Kim. This book provides empirical research backing the effectiveness of DevOps practices and the importance of culture and automation in achieving high performance in technology organizations, supporting the chapter's discussion on continuous practices in DevOps, DevSecOps, and **site reliability engineering (SRE)**.

- *Artificial Intelligence: A Guide for Thinking Humans* by Melanie Mitchell. This book provides a deep dive into the capabilities and limitations of current AI technologies, including their applications in various fields such as software development. It offers a foundational understanding of how AI and ML can be leveraged for continuous testing, quality, security, and feedback, which aligns well with the themes discussed in *Chapter 8*.

- *Continuous Delivery: Reliable Software Releases through Build, Test, and Deployment Automation* by Jez Humble and David Farley. This book focuses on the principles of CD, which are closely related to the continuous testing and feedback mechanisms discussed in *Chapter 3*. It provides a framework that supports the chapter's emphasis on learning from past experiences to improve future practices.

- *Continuous Security Testing: Integrating Protection into DevOps* by Kualitatem. This blog post explores the integration of continuous security measures into the DevOps pipeline, which supports the chapter's discussion on continuous security practices within DevSecOps.

- *Data Science for Business: What You Need to Know about Data Mining and Data-Analytic Thinking* by Foster Provost and Tom Fawcett. This book offers insights into how data-driven methods can be applied to business challenges, including software development and operations. It covers foundational data science principles that are crucial for understanding how to effectively implement AI/ML in continuous processes.

- *DevOps for Dummies* by Emily Freeman. Aimed at beginners and early adopters of DevOps, this book describes DevOps as an engineering culture of collaboration, ownership, and learning, supporting the chapter's emphasis on continuous improvement within DevOps practices.

- *Engineering DevOps: From Chaos to Continuous Improvement* by Marc Hornbeek. This book is an excellent reference guide for anyone looking to implement or improve DevOps practices. It covers the necessary engineering practices to achieve continuous improvement and beyond, making it highly relevant to the attached chapter's focus on continuous learning and improvement.

- *Lean Analytics: Use Data to Build a Better Startup Faster* by Alistair Croll and Benjamin Yoskovitz. This book provides a comprehensive guide on how to use data effectively to measure and drive business success. It aligns well with the themes of *Chapter 12*, particularly in the context of measuring progress and outcomes in organizational transformation.

- *Machine Learning Yearning* by Andrew Ng. Written by one of the pioneers in the field, this book focuses on how to structure **machine learning** (**ML**) projects. Andrew Ng provides insights into the practical aspects of applying ML to complex systems, which can be directly applied to enhancing software development processes as outlined in this book.

- *Project to Product: How to Survive and Thrive in the Age of Digital Disruption with the Flow Framework* by Mik Kersten. Kersten's book introduces the Flow Framework, a new way of building an infrastructure for innovation that focuses on the flow of value from an organization to its customers. This framework aligns well with the strategic planning and roadmap development for implementing continuous practices, as discussed in *Chapter 10*.

- *Site Reliability Engineering: How Google Runs Production Systems* by Niall Richard Murphy, Betsy Beyer, Chris Jones, and Jennifer Petoff. This book introduces the principles and practices of SRE, offering insights into how Google ensures the reliability and scalability of its systems, which complements the chapter's focus on continuous operations practices known as SRE.

- *The DevOps Handbook: How to Create World-Class Agility, Reliability, & Security in Technology Organizations* by Gene Kim, Patrick Debois, John Willis, and Jez Humble. This book provides insights into integrating continuous testing, security, and feedback into DevOps practices, which aligns with the experiences and lessons shared in *Chapter 3* regarding the evolution and practical application of these concepts.

- *The Importance of Continuous Feedback in Software Testing* by GeeksforGeeks. This article highlights the role of continuous feedback in software testing, aligning with the chapter's emphasis on continuous feedback mechanisms within Agile, DevOps, and SRE methodologies.

- *The Importance of Continuous Testing in Software Development* by Synopsys. This article discusses the strategic implementation of continuous testing within DevOps, aligning with the chapter's focus on continuous testing as part of digital transformations in software development.

- *The Lean Startup: How Today's Entrepreneurs Use Continuous Innovation to Create Radically Successful Businesses* by Eric Ries. Eric Ries introduces concepts such as the build-measure-learn feedback loop, which is crucial for measuring progress and outcomes in any innovative process, including those involving continuous testing, quality, security, and feedback.

- *The Phoenix Project: A Novel about IT, DevOps, and Helping Your Business Win* by Gene Kim, Kevin Behr, and George Spafford. This novel provides insights into the DevOps philosophy and its impact on improving business outcomes. It emphasizes the importance of collaboration, automation, and learning from failures, aligning with the attached chapter's themes.

- *How to Measure Anything: Finding the Value of 'Intangibles' in Business* by Douglas W. Hubbard. This book challenges the notion that some things are immeasurable, and it provides methods and best practices for quantifying different types of risk and value in a business context, directly supporting the measurement strategies discussed in *Chapter 12*.

Internet references

- *AI in Plain English*: `https://medium.com/ai-in-plain-english`
  This Medium publication breaks down complex AI concepts into understandable articles, often focusing on practical applications. It includes discussions on how AI can transform industries, including software development, by automating tasks, improving quality, and enhancing security measures.

- *Blog Post on DevOps.com by Marc Hornbeek*
  Marc Hornbeek regularly contributes articles to `DevOps.com`, where he discusses various aspects of DevOps practices, including continuous testing, quality, security, and feedback. These articles provide practical insights and complement the discussions in this book.

- *Cloud Native Computing Foundation (CNCF) Blog*: `https://www.cncf.io/blog/`
  These insights and updates on cloud-native technologies and practices support the integration of digital technology into business processes as outlined in the digital transformation strategies in the chapter.

- *Comparing SRE and DevOps: What Are the Differences?* StrongDM.
  This article provides a detailed comparison between SRE and DevOps, highlighting their origins, principles, and practices. It supports the attached chapter's discussion on various ways teams can learn and improve through experimentation, handling failures, and applying chaos engineering.

- *Continuous Feedback Use Case: Stakeholder Feedback on Requirements*
  Gathering and incorporating feedback from stakeholders on the defined requirements and quality criteria. Ensures that the project aligns with stakeholder expectations and business objectives.

- *Continuous Security Testing: Integrating Protection into DevOps*. Kualitatem.
  This blog post explores the integration of continuous security measures into the DevOps pipeline, which supports the chapter's discussion on continuous security practices within DevSecOps.

- *DevOps Institute*: `https://devopsinstitute.com/`
  The DevOps Institute offers resources, certifications, and educational materials on DevOps practices, including continuous testing, quality, security, and feedback. It's a valuable resource for understanding the latest trends and advancements in DevOps and for finding strategies to implement these practices effectively.

- *DevOps.com*: `https://devops.com/`
  A leading source for articles, news, and analysis on DevOps, including topics on continuous testing, security, and feedback. This site provides practical insights and case studies relevant to the concepts discussed in *Chapter 4*.

- *DZone*: `https://dzone.com/`
  DZone offers a wide range of articles and tutorials on software development and various aspects of DevOps, including continuous testing and security. It's a valuable resource for understanding the application of transformation goals in real-world scenarios.

- *Google AI Blog*: `https://ai.googleblog.com/`
  This blog provides updates and insights from Google's AI research and projects. It is a great resource for staying informed about the latest advancements in AI and ML technologies, including their applications in software engineering and DevOps practices.

- *Google Analytics*: `https://analytics.google.com`
  Google Analytics provides tools for measuring website and campaign performance, which can be analogous to measuring project outcomes and progress. It offers insights into effectively tracking and interpreting user data, relevant to the digital aspects of business transformations.

- *InfoQ*: `https://www.infoq.com/`
  InfoQ provides updates and deep dives into the latest trends and innovations in software development, including CD and DevOps practices. It helps readers stay informed about the latest strategies and technologies that impact transformation goals.

- *Integrating Security into DevOps: Bridging the Gap Between Speed and Security.*
  Security Boulevard.
  This article explores the integration of security practices within DevOps workflows, supporting the chapter's discussion on the evolution of DevSecOps and the practical challenges of embedding security throughout the development life cycle.

- *Interview on CloudNativeNow.com with Marc Hornbeek*
  In interviews, Marc Hornbeek often discusses his experiences and insights into DevOps and continuous practices, providing real-world examples and strategies that support the content of this book.

- *Kissmetrics Blog*: `https://blog.kissmetrics.com/`
  This blog offers articles on analytics, marketing, and testing, providing valuable insights into how data can drive decision-making and improve business processes, aligning with the continuous improvement and measurement themes of *Chapter 12*.

- *Learning from Production: How Real-Time Monitoring and Feedback Can Improve Software Development.* DevOps.com.
  This article discusses the importance of real-time monitoring and feedback in software development, which aligns with the chapter's focus on learning from real-world experiences to refine testing and security practices.

- *Security Boulevard*: `https://securityboulevard.com/`
  Focuses on security in the IT sector, offering articles and expert opinions that align with the continuous security aspects of the Capability Maturity Model discussed in the chapter.

- *SRE vs. DevOps: What's the difference?* TechTarget.
  This article explores the differences between SRE and DevOps, emphasizing continuous improvement through appropriate monitoring tools and practices. It aligns with the attached chapter's emphasis on continuous learning and improvement in software development and operations.

- *Tableau*: https://www.tableau.com/
  Tableau offers powerful data visualization tools that help organizations turn large amounts of data into actionable insights. This is particularly useful for the visualization of progress and outcome metrics as discussed in *Chapter 12*.

- *The Importance of Continuous Feedback in Software Testing.* GeeksforGeeks.
  This article highlights the role of continuous feedback in software testing, aligning with the chapter's emphasis on continuous feedback mechanisms within Agile, DevOps, and SRE methodologies.

- *The New Stack*: https://thenewstack.io/
  This focuses on the analysis of trends in software development, including DevOps and cloud-native technologies. The New Stack offers articles that explore how emerging trends are being implemented in the industry.

- *The Role of Feedback in Agile Development: How Continuous Feedback Drives Improvement.* Agile Alliance.
  This article highlights the role of continuous feedback in agile development processes, which complements the chapter's emphasis on using feedback to enhance quality and security in software projects.

- *Towards Data Science*: https://towardsdatascience.com/
  A platform where data science professionals and enthusiasts share insights and tutorials, including applications of AI and ML in various industries. Articles on this site often explore innovative ways to apply AI/ML to software development, making it a valuable resource for understanding practical implementations similar to those discussed in *Chapter 8*.

- *Transform Your DevOps, DevSecOps and SRE to Cloud Native.* CloudNativeNow.com.
  This article discusses the strategic move of embracing cloud-native principles in DevOps, DevSecOps, and SRE practices. It highlights the benefits of increased agility, scalability, and resilience, supporting the attached chapter's focus on using technology changes as opportunities for advancement.

- *Webinar Featuring Marc Hornbeek*
  Marc Hornbeek has participated in several webinars focusing on DevOps, continuous testing, and security. These webinars often book themes.

- *What is Continuous Testing in DevOps? (Strategy + Tools).* TestRail.
  This article discusses the strategic implementation of continuous testing within DevOps, aligning with the chapter's focus on continuous testing as part of digital transformations in software development.

Index

packtpub.com

Subscribe to our online digital library for full access to over 7,000 books and videos, as well as industry leading tools to help you plan your personal development and advance your career. For more information, please visit our website.

Why subscribe?

- Spend less time learning and more time coding with practical eBooks and Videos from over 4,000 industry professionals

- Improve your learning with Skill Plans built especially for you

- Get a free eBook or video every month

- Fully searchable for easy access to vital information

- Copy and paste, print, and bookmark content

Did you know that Packt offers eBook versions of every book published, with PDF and ePub files available? You can upgrade to the eBook version at packtpub.com and as a print book customer, you are entitled to a discount on the eBook copy. Get in touch with us at customercare@packtpub.com for more details.

At www.packtpub.com, you can also read a collection of free technical articles, sign up for a range of free newsletters, and receive exclusive discounts and offers on Packt books and eBooks.

Other Books You May Enjoy

If you enjoyed this book, you may be interested in these other books by Packt:

Efficient Cloud FinOps

Alfonso San Miguel Sánchez, Danny Obando García

ISBN: 978-1-80512-257-9

- Examine challenges in cloud adoption and cost optimization
- Gain insight into the integration of FinOps within organizations
- Explore the synergies between FinOps and DevOps, IaC, and change management
- Leverage tools such as Azure Advisor, AWS CUDOS, and GCP cost reports
- Estimate and optimize costs using cloud services key features and best practices
- Implement cost dashboards and reports to improve visibility and control
- Understand FinOps roles and processes crucial for organizational success
- Apply FinOps through real-life examples and multicloud architectures

Modern DevOps Practices

Gaurav Agarwal

ISBN: 978-1-80512-182-4

- Explore modern DevOps practices with Git and GitOps
- Master container fundamentals with Docker and Kubernetes
- Become well versed in AWS ECS, Google Cloud Run, and Knative
- Discover how to efficiently build and manage secure Docker images
- Understand continuous integration with Jenkins on Kubernetes and GitHub Actions
- Get to grips with using Argo CD for continuous deployment and delivery
- Manage immutable infrastructure on the cloud with Packer, Terraform, and Ansible
- Operate container applications in production using Istio and learn about AI in DevOps

Packt is searching for authors like you

If you're interested in becoming an author for Packt, please visit authors.packtpub.com and apply today. We have worked with thousands of developers and tech professionals, just like you, to help them share their insight with the global tech community. You can make a general application, apply for a specific hot topic that we are recruiting an author for, or submit your own idea.

Share Your Thoughts

Now you've finished *Continuous Testing, Quality, Security, and Feedback*, we'd love to hear your thoughts! Scan the QR code below to go straight to the Amazon review page for this book and share your feedback or leave a review on the site that you purchased it from.

https://packt.link/r/1835462243

Your review is important to us and the tech community and will help us make sure we're delivering excellent quality content.

Download a free PDF copy of this book

Thanks for purchasing this book!

Do you like to read on the go but are unable to carry your print books everywhere?

Is your eBook purchase not compatible with the device of your choice?

Don't worry, now with every Packt book you get a DRM-free PDF version of that book at no cost.

Read anywhere, any place, on any device. Search, copy, and paste code from your favorite technical books directly into your application.

The perks don't stop there, you can get exclusive access to discounts, newsletters, and great free content in your inbox daily

Follow these simple steps to get the benefits:

1. Scan the QR code or visit the link below

https://packt.link/free-ebook/9781835462249

2. Submit your proof of purchase
3. That's it! We'll send your free PDF and other benefits to your email directly